# Beekeeping

## for dummies®

A Wiley Brand

# Beekeeping

4th Edition

**by Howland Blackiston**

FOREWORD BY **Dewey M. Caron**

**Beekeeping For Dummies®, 4th Edition**

Published by:
**John Wiley & Sons, Inc.**
111 River Street
Hoboken, NJ 07030-5774
www.wiley.com

Copyright © 2017 by John Wiley & Sons, Inc., Hoboken, New Jersey

Published simultaneously in Canada

For general information on our other products and services, please contact our Customer Care Department within the U.S. at 877-762-2974, outside the U.S. at 317-572-3993, or fax 317-572-4002. For technical support, please visit https://hub.wiley.com/community/support/dummies.

Wiley publishes in a variety of print and electronic formats and by print-on-demand. Some material included with standard print versions of this book may not be included in e-books or in print-on-demand. If this book refers to media such as a CD or DVD that is not included in the version you purchased, you may download this material at http://booksupport.wiley.com. For more information about Wiley products, visit www.wiley.com.

Library of Congress Control Number: 2016960491

ISBN 978-1-119-31006-8 (pbk); ISBN 978-1-119-31102-7 (ebk); ISBN 978-1-119-31103-4 (ebk)

Manufactured in the United States of America

10  9  8  7  6  5  4  3  2  1

# Foreword

In his foreword to the third edition of *Beekeeping For Dummies,* Connecticut author and beekeeper Ed Weiss wrote, "This book is written for all those folks who have happily decided to have our wonderful friend, the honeybees, as their own." Ed concludes that *Beekeeping For Dummies* answers beekeepers' perpetual "How do I do it?" question. That continues to hold true for the updated and expanded fourth edition.

The learning curve to successfully starting and continuing to maintain beehives can be steep. *Beekeeping For Dummies* flattens the curve in covering all aspects of keeping bees. Anyone interested in learning about the basics of beekeeping — from figuring out what equipment is required to what to do after the bees are installed to harvesting honey to helping colonies with disease and pest control — will find it all here. This book is ideal for everyone, from those with a blossoming interest in bees to those who have already started. It eliminates the uncertainties — our dumbness — of bees and beekeeping.

*Dummies* author Howland Blackiston is no stranger to honey bees. He has been managing bees, starting with an observation hive, for more than 30 years. He speaks authoritatively of his firsthand experiences. He offers his recommendations gently and honestly, based on his successes. And he encourages new beekeepers to experiment with their own techniques when traditional solutions don't work as expected.

Howland is an accomplished honey wine (mead) connoisseur, makes beautiful beeswax candles, and has sampled honeys from around the world. He is a conscientious bee mentor for "newbees." As you read, you'll find him to be a patient and thorough teacher of beekeeping.

The fourth edition of *Beekeeping For Dummies* has been expanded, in response to reader demand, to include information on Top Bar hives. It still includes those items readers have come to expect from earlier editions. Icons highlight the text with tips, reminders, and warnings as well as call attention to specialty concerns that urban beekeepers face. And all beekeepers would do well to pay attention to Howland's explanations of how to be all-natural, pointed out by another icon. Brand-new in this fourth edition is information on Top Bar hives.

In addition to money-saving coupons, there are directions to find online videos (in other words, to see, as well as read about, techniques as you set up a new hive or prepare to harvest honey and carry out the honey extracting process). You can also find a Dummies Cheat Sheet online to keep up with your seasonality practices.

The book promises and delivers step-by-step, easy-to-understand instructions to becoming organized and beginning a bee colony. Seasonal chapters explain what to do and when to do it. This is a great book to start beekeeping. However, readers will find much more as they refer back to the information or consult the additional information found online. I wholeheartedly recommend that this book be part of your beekeeping bookshelf. I expect it will become a regular read.

Dewey M. Caron
Emeritus Professor University of Delaware
Retired after 40 years of bee extension, research, and University teaching
Still interacting with bee associations, bee short courses, and outreach

# Contents at a Glance

# Table of Contents

# Introduction

Keeping honey bees is a unique and immensely rewarding hobby. If you have an interest in nature, you'll deeply appreciate the wonderful world that beekeeping opens up to you. If you're a gardener, you'll treasure the extra bounty that pollinating bees bring to your fruits, flowers, and vegetables. If you're a foodie, you will celebrate your own honey harvest. In short, you'll be captivated by these remarkable little creatures in the same way others have been captivated for thousands of years.

Becoming a beekeeper is easy and safe — it's a great hobby for the entire family. All you need is a little bit of guidance to get started. And that's exactly what this book is for. I provide you with a step-by-step approach for successful backyard beekeeping — follow it closely, and you can have a lifetime of enjoyment with your bees.

## About This Book

This book is a reference, not a lecture. You certainly don't have to read it from beginning to end unless you want to. I organized the chapters in a logical fashion with sensitivity to the beekeeper's calendar of events. I include lots of great photographs and illustrations (each, I hope, is worth a thousand words) and lots of practical advice and suggestions.

Because Langstroth hives are far and away the most widely used type of hive in the world, most of the content is written with the assumption that you will start your adventure using a Langstroth hive. But since Top Bar hives (and Kenyan Top Bar hives in particular) have become increasingly popular, this fourth edition includes information of particular interest to you Top Bar beekeepers.

I guarantee you will discover all sorts of new information and helpful tips. Just take a look at the sorts of things I've included. This book

>> Lets you explore the many benefits of beekeeping and helps you decide whether beekeeping is the right hobby for you.

>> Gives you some insight into a day in the life of the honey bee. You find out about the queen, the workers, and the drones, and the roles each plays in the colony. It explains basic bee anatomy and introduces you to other stinging insects.

>> Deals with any apprehensions you may have about beekeeping (stings, neighbors' funny looks, local restrictions, environmental considerations, costs, and time and commitment).

>> Helps you decide where you should locate your hive and how you can get started.

>> Introduces you to five different hive types that are the most popular among backyard beekeepers. You learn the benefits and drawbacks associated with each of these hives.

>> Shows the basic tools and equipment you need and how to assemble a hive. You find out about really cool gadgets and weird and wonderful accessories.

>> Helps you decide the kind of honey bee to raise, and when and how to order your bees. You also find out what to do the day your "girls" arrive and how to successfully transfer them to their new home.

>> Explains how and when to go about approaching and opening up a colony of bees.

>> Helps you understand exactly what you're looking for every time you inspect a colony. I include the specific tasks that are unique to the weeks immediately following the arrival of your bees as well as throughout the season.

>> Discusses the tasks a beekeeper must perform year-round to maintain a healthy colony. Use it as a checklist of seasonal activities that you can refer back to year after year. There's a neat "Beekeeper's Calendar" that's keyed to different climates. Use this to identify the tasks you should do and when.

>> Shows you how to anticipate the most common problems. Find out what to do if your hive swarms or simply packs up and leaves. Discover how to recognize problems with brood production and your precious queen.

>> Provides the latest information about topics that are all the buzz in the media: colony collapse disorder (CCD), backyard pollination, nutritional value of honey, and Africanized bees. Find out what you can do to help save the honey bees.

>> Takes a detailed look at bee illnesses. Learn what medications you can use to keep your bees healthy and productive, year after year. Discover alternative natural approaches to keeping bees healthy that avoid or minimize the use of any chemicals.

>> Teaches you the basics of raising your own queen bees for fun and profit.

>> Gives you a yummy chapter 100 percent about honey. Find out about its influence throughout history, what it's composed of, the many different kinds and characteristics of honey, and how honey is used for health and wellness.

>> Gives you a step-by-step approach for harvesting, bottling, and marketing your honey.

I also include some back-of-book materials, including helpful bee-related resources: websites, journals, suppliers, and beekeeping associations. I also give you a glossary of bee and beekeeping terms that you can use as a handy quick reference and some useful templates for creating your own beekeeping checklists and logs. Finally, there are some special discount coupons from major vendors that you can take advantage of for managing your hives, purchasing equipment, and subscribing to one of the leading bee journals.

# Foolish Assumptions

If you've never kept bees, this book has all the information you need to get started in beekeeping. I assume that you have no prior knowledge of the equipment, tools, and techniques.

However, if you've been a beekeeper for a while, this book is a terrific resource for you, too. You'll find new ideas on how to keep your bees healthier and more productive, as well as natural alternatives to traditional medication approaches. And if you're a city dweller, you'll benefit from the special hints and guidelines that are unique to urban beekeeping. You'll appreciate the way the book has been organized for easy and ongoing reference. I include a whole lot of "tricks of the trade." In short, this book is for just about anyone who's fallen in love with the bountiful honey bee.

# Icons Used in This Book

Peppered throughout this book are helpful icons that present special types of information to enhance your reading experience and make you a stellar beekeeper.

**TIP**

Think of these tips as words of wisdom that — when applied — can make your beekeeping experience more pleasant and fulfilling!

**WARNING**

These warnings alert you to potential beekeeping boo-boos that could make your experiences unpleasant and/or your little winged friends unhappy or downright miserable. Take them to heart!

**REMEMBER**

I use this icon to point out things that need to be so ingrained in your beekeeping consciousness that they become habits. Keep these points at the forefront of your mind when caring for your bees.

**URBAN**

Urban beekeeping involves special considerations and techniques. For those of you keeping bees in a city environment, these hints and suggestions will come in very handy.

**ALL NATURAL**

The trend these days is to be very judicial about the use of chemicals in the hive. Use this icon to learn about natural, nonchemical methods for tending to your colony's health and otherwise dealing with problems.

# Beyond the Book

There is much more information available from your author, and from the *For Dummies* brand, for your learning pleasure. Check out these resources to find out more about the art of beekeeping:

>> The Cheat Sheet gives you quick reference tasks that you need to complete at various times of the year. Find out what you need to do in summer and winter to care for and protect your bees. To get this Cheat Sheet, simply go to www.dummies.com and search for "Beekeeping For Dummies Cheat Sheet" in the Search box.

>> *For Dummies* online videos are available at www.dummies.com/go/beekeepingfd4e. You can see everything from installing your bees to lighting a smoker. Watching me perform these tasks is an invaluable help to new and seasoned beekeepers.

>> And, although this book includes information about different beehives, I give you much more information in *Building Beehives For Dummies* (Wiley).

# Where to Go from Here

You can start anywhere with *For Dummies* books, but there's a logic to beginning at the beginning. If that's not in your personality, consider starting with Chapter 5 to see what equipment you'll need to get started. Then move over to Chapter 6 to determine what kind of honey bee is right for you and what to do the day your "girls" arrive at your home.

You may have some apprehension about working around bees, such as stings and whether your neighbors will be comfortable with your new hobby. Check out Chapter 3 to get some ideas on how to win over your neighbors with jars of delicious honey.

Colony collapse disorder has been in the news over the last few years. Chapter 11 takes an in-depth look at this problem and answers common questions. Chapter 8 gives you info and advice about inspecting your hives and your bees, and Chapters 12 and 13 tell you what to do if you find mites or other potentially fatal problems with your bees.

Chapter 15 is all about honey. Sweet, sticky, golden honey. You learn about the health benefits as well as a little about the history and the medicinal benefits. You find out how to gather and process your honey in Chapters 16 and 17. And after all the work is done and you have pounds of the sweet stuff, Chapters 18 and 20 give you directions for everything from brewing your own mead (honey wine) to making your own lip balm, along with lots of yummy honey-inspired recipes.

My advice is to not hurry through this book. There's a lot of information here, and all of it will help you be a better beekeeper. Whether you're in the country, in a subdivision, or you're planning on an urban rooftop beehive, there's something here for you.

# 1

# Taking Flight with Beekeeping

Discover the role honey bees have played in history, and find out about the many benefits of beekeeping today. Understand the honey bees' vital role in nature and what they contribute to our agricultural economy.

Decide whether becoming a beekeeper is the right decision for you. How much work is involved? Do you have what it takes? What will your neighbors say?

See what makes honey bees tick. Understand how they communicate with each other, and find out about their different roles and responsibilities as members of a fascinating society.

Find out about other stinging insects that are often confused with the gentle honey bee and as a result give them a bad rap.

# Chapter **1**

# To Bee, or Not to Bee?

've been keeping bees in my backyard since 1983, and I have a confession to make — I really love my bees. That may sound weird to you if you aren't a bee-keeper (yet!), but virtually all individuals who keep bees will tell you the same thing and speak with affection about "their girls." They impatiently await their next opportunity to visit their hives. They experience a true emotional loss when their bees don't make it through a bad winter. Beekeepers, without a doubt, develop a special bond with their bees.

Since becoming a backyard beekeeper, I've grown to deeply admire the remarkable qualities of these endearing creatures. As a gardener, I've witnessed firsthand the dramatic contribution they provide to flowering plants of all kinds. With honey bees in my garden, its bounty has increased by leaps and bounds. And then there's that wonderful bonus that they generously give me: a yearly harvest of sweet liquid gold.

After you get to know more about bees' value and remarkable social skills, you'll fall in love with them, too. They're simply wonderful little creatures. Interacting with them is a joy and a privilege. People who love nature in its purest form will love bees and beekeeping.

That being said, in this chapter, I help you better understand the remarkable and bountiful little honey bee by looking at its history and the value it brings to our lives. I also discuss the benefits of beekeeping and why you should consider it as a hobby — or even a small business venture. This chapter outlines the benefits of keeping bees, offers an overview of what is required to keep bees, and explains the various approaches you can take to keep your bees healthy.

## THE PREHISTORIC BEE

Bees have been around for a long, long time, gathering nectar and pollinating flowers. They haven't changed much since the time of the dinosaurs. The insect shown in the following figure is definitely recognizable as a bee. It was caught in a flow of pine sap 30 million to 40 million years ago and is forever preserved in amber.

*Courtesy of Mario Espinola,* www.espd.com

# Discovering the Benefits of Beekeeping

Why has mankind been so interested in beekeeping over the centuries? I'm sure that the first motivator was *honey.* After all, for many years and long before cane sugar, honey was the primary sweetener in use. I'm also sure that honey remains the principal draw for many backyard beekeepers. Chapters 15 through 17 explain about all the different kinds of honey, its culinary attraction, and how to produce, harvest, and market your honey.

But the sweet reward is by no means the only reason folks are attracted to beekeeping. Since the 18th century, agriculture has recognized the value of pollination by bees. Without the bees' help, many commercial crops would suffer serious consequences. More on that later. Even backyard beekeepers witness dramatic improvements in their garden: more and larger fruits, flowers, and vegetables. A hive or two in the garden makes a big difference in your success as a gardener.

The rewards of beekeeping extend beyond honey and pollination. Bees produce other products that can be harvested and put to good use, including beeswax, propolis, and royal jelly. Even the pollen they bring back to the hive can be harvested (it's rich in protein and makes a healthy food supplement in our own diets). Chapter 18 includes a number of practical things you can do with beeswax and propolis.

## Harvesting liquid gold: Honey

The prospect of harvesting honey is certainly a strong attraction for new beekeepers. There's something magical about bottling your own honey. And I can assure you that no other honey tastes as good as the honey made by your own bees. Delicious! Learn all about honey varieties, tasting, and pairing with food in Chapter 15. And be sure to have a look at Chapter 20, where I list ten of my favorite recipes for cooking with honey.

How much honey can you expect? The answer to that question varies depending on the weather, rainfall, and location and strength of your colony. But producing 60 to 80 pounds or more of surplus honey isn't unusual for a single strong colony. Chapters 15 through 17 provide plenty of useful information on the kinds of honey you can harvest from your bees and how to go about it. Also included are some suggestions on how you can go about selling your honey — how this hobby can boast a profitable return on investment!

## Bees as pollinators: Their vital role to our food supply

Any gardener recognizes the value of pollinating insects. Various insects perform an essential service in the production of seed and fruit. The survival of plants depends on pollination. You may not have thought much about the role honey bees play in our everyday food supply. It is estimated that in North America around 30 percent of the food we consume is produced from bee-pollinated plants. Bees also pollinate crops, such as clover and alfalfa that cattle feed on, making bees important to our production and consumption of meat and dairy. The value of pollination by bees is estimated at around $16 billion in the United States alone.

These are more than interesting facts; these are realities with devastating consequences if bees were to disappear. And sadly, the health of honey bees has been compromised in recent years (see the later section "Being part of the bigger picture: Save the bees!"). Indeed a spring without bees could endanger our food supply and impact our economy. It's a story that has become headline news in the media.

## HONEYBEE OR HONEY BEE?

This is a "tomato/tomahto" issue. The British adhere to their use of the one word: "honeybee." The Entomological Society of America, however, prefers to use two words "honey bee." So that's the version I use in this book. Here's the society's rationale: The honey bee is a true bee, like a house fly is a true fly, and thus should be two words. A dragonfly, on the other hand, is not a fly; hence, it is one word. *Tip:* Spell it both ways when web surfing. That way, you'll cover all bases and hit all the sites!

I've witnessed the miracle in my own garden: more and larger flowers, fruits, and vegetables — all the result of more efficient pollination by bees. After seeing my results, a friend who tends an imposing vegetable garden begged me to place a couple of hives on her property. I did, and she too is thrilled. She rewards me with a never-ending bounty of fruits and vegetables. And I pay my land-rent by providing her with 20 pounds of honey every year. Not a bad barter all around.

## WHY BEES MAKE GREAT POLLINATORS

About 100 crops in the United States depend on bees for pollination. Why is the honey bee such an effective pollinator? Because she's uniquely adapted to the task. Here are several examples:

- The honey bee's anatomy is well suited for carrying pollen. Her body and legs are covered with branched hairs that catch and hold pollen grains. The bee's hind legs contain *pollen baskets* that the bee uses for transporting pollen, a major source of food, back to the hive. If the bee brushes against the stigma (female part) of the next flower she visits, some of the pollen grains brush off, and the act of cross-pollination is accomplished.

- Most other insects lie dormant all winter and in spring emerge only in small numbers until increasing generations have rebuilt the population of the species. Not the honey bee. Its hive is perennial. The honey bee overwinters with large numbers of bees feeding on stored honey and pollen. Early in the spring, the queen begins laying eggs, and the already large population explodes. When flowers begin to bloom, each hive has tens of thousands of bees to carry out pollination activities. By midsummer, an individual hive contains upward of 60,000 bees.

- The honey bee has a unique habit that's of great value as a pollinator. It tends to forage on blooms of the same kind, as long as they're flowering. In other words, rather than traveling from one flower type to another, honey bees are flower constant. This focus makes for particularly effective pollination. It also means that the honey they produce from the nectar of a specific flower takes on the unique flavor characteristics of that flower — that's how we get specific honey flavors, such as orange blossom honey, buckwheat honey, blueberry honey, lavender honey, and so on (see Chapter 15 for a lesson in how to taste and evaluate honey varieties).

- The honey bee is one of the only pollinating insects that can be introduced to a garden at the gardener's will. You can garden on a hit-or-miss basis and hope that enough wild bees are out there to achieve adequate pollination — or you can take positive steps and nestle a colony of honey bees in a corner of your garden. Some commercial beekeepers make their living by renting colonies of honey bees to farmers who depend on bee pollination to raise more bountiful harvests. Known as *migratory beekeepers*, they haul hundreds of hives across the country, following the various agricultural blooms — to California for almond pollination in February, to the apple orchards in Washington in April, to Maine in May for blueberry pollination, and so on.

## Being part of the bigger picture: Save the bees!

The facts that keeping a hive in the backyard dramatically improves pollination and rewards you with a delicious honey harvest are by themselves good enough reasons to keep bees. But today, the value of keeping bees goes beyond the obvious. In many areas, millions of colonies of wild (or *feral*) honey bees have been wiped out by urbanization, pesticides, parasitic mites, and a recent phenomenon called *colony collapse disorder* (otherwise known as CCD; see Chapter 11 for more information). Collectively, these challenges are devastating the honey bee population. Many gardeners have asked me why they now see fewer and fewer honey bees in their gardens. It's because of the dramatic decrease in our honey bee population. Backyard beekeeping has become vital in our efforts to reestablish new colonies of honey bees and offset the natural decrease in pollination by wild honey bees. I know of many folks who have started beekeeping just to help rebuild the honey bee population.

## Getting an education: And passing it on!

As a beekeeper, you continually discover new things about nature, bees, and their remarkable social behavior. Just about any school, nature center, garden club, or youth organization would be delighted for you (as a beekeeper) to share your knowledge. Each year I make the rounds with my slide show and props, sharing the miracle of honey bees with communities near and far. On many occasions, local teachers and students have visited my house for an on-site workshop. I've opened the hive and given each wide-eyed student a close-up look at bees at work. Spreading the word to others about the value these little creatures bring to all of us is great fun. You're planting a seed for our next generation of beekeepers. After all, a grade-school presentation on beekeeping is what aroused my interest in honey bees.

## BEE HUNTERS, GATHERERS, AND CULTIVATORS

An early cave painting in Biscorp, Spain, circa 6000 BC, shows early Spaniards hunting for and harvesting wild honey (see the accompanying figure). In centuries past, honey was a treasured and sacred commodity. It was used as money and praised as the nectar of the gods. Methods of beekeeping remained relatively unchanged until 1852 with the introduction of today's "modern" removable-frame hive, also known as the Langstroth hive. (See Chapter 4 for more information about Langstroth and other kinds of beehives.)

*Courtesy of Howland Blackiston*

# Improving your health: Bee therapies and stress relief

Although I can't point to any scientific studies to confirm it, I honestly believe that tending honey bees reduces stress. Working with my bees is so calming and almost magical. I am at one with nature, and whatever problems may have been on my mind tend to evaporate. There's something about being out there on a lovely warm day, the intense focus of exploring the wonders of the hive, and hearing that gentle hum of contented bees — it instantly puts me at ease, melting away whatever day-to-day stresses I may find creeping into my life.

Any health food store proprietor can tell you the benefits of the bees' products. Honey, pollen, royal jelly, and propolis have been a part of healthful remedies for centuries. Honey and propolis have significant antibacterial qualities. Royal jelly is loaded with B vitamins and is widely used overseas as a dietary and fertility stimulant. Pollen is high in protein and can be used as a homeopathic remedy for seasonal pollen allergies (see the sidebar "Bee pollen, honey, and allergy relief" in this chapter).

## BEE POLLEN, HONEY, AND ALLERGY RELIEF

Pollen is one of the richest and purest of natural foods, consisting of up to 35 percent protein and 10 percent sugars, carbohydrates, enzymes, minerals, and vitamins A (carotenes), B1 (thiamin), B2 (riboflavin), B3 (nicotinic acid), B5 (panothenic acid), C (ascorbic acid), H (biotin), and R (rutine).

Here's the really neat part: Ingesting small amounts of pollen every day can actually help reduce the symptoms of pollen-related allergies — sort of a homeopathic way of inoculating yourself.

Of course you can harvest pollen from your bees and sprinkle a small amount on your breakfast cereal or in yogurt (as you might do with wheat germ). But you don't really need to harvest the pollen itself. That's because raw, natural honey contains pollen. Pollen's benefits are realized every time you take a tablespoon of honey. Eating local honey every day can relieve the symptoms of pollen-related allergies if the honey is harvested from within a 50-mile radius of where you live or from an area where the vegetation is similar to what grows in your community. Now that you have your own bees, that isn't a problem. Allergy relief is only a sweet tablespoon away!

*Apitherapy* is the use of bee products for treating health disorders. Even the bees' venom plays an important role here — in bee-sting therapy. Venom is administered with success to patients who suffer from arthritis and other inflammatory/medical conditions. This entire area has become a science in itself and has been practiced for thousands of years in Asia, Africa, and Europe. An interesting book on apitherapy is *Bee Products — Properties, Applications and Apitherapy: Proceedings of an International Conference Held in Tel Aviv, Israel, May 26–30, 1996,* published by Kluwer Academic Publishers.

# Determining Your Beekeeping Potential

How do you know whether you'd make a good beekeeper? Is beekeeping the right hobby for you? Here are a few things worth considering as you ponder these issues.

## Environmental considerations

Unless you live on a glacier or on the frozen tundra of Siberia, you probably can keep bees. Bees are remarkable creatures that do just fine in a wide range of climates. Beekeepers can be found in areas with long, cold winters; in tropical rain forests; and in nearly every geographic region in between. If flowers bloom in your part of the world, you can keep bees.

How about space requirements? You don't need much. I know many beekeepers in the heart of Manhattan. They have a hive or two on their rooftops or terraces. Keep in mind that bees travel one to two miles from the hive to gather pollen and nectar. They'll forage an area as large as 6,000 acres, doing their thing. So the only space that you need is enough to accommodate the hive itself.

See Chapter 3 for more specific information on where to locate your bees, in either urban or suburban situations.

## Zoning and legal restrictions

Most communities are quite tolerant of beekeepers, but some have local ordinances that prohibit beekeeping or restrict the number of hives you can have. Some communities let you keep bees but ask that you register your hives with the local government. Check with local bee clubs, your town hall, your local zoning board, or your state's Department of Agriculture (bee/pollinating insects division) to find out about what's okay in your neighborhood.

Obviously you want to practice a good-neighbor policy so folks in your community don't feel threatened by your unique new hobby. See Chapter 3 for more information on the kinds of things you can do to prevent neighbors from getting nervous.

# Costs and equipment

What does it cost to become a beekeeper? All in all, beekeeping isn't a very expensive hobby. You can figure on investing about $300 to $400 for a start-up hive kit, equipment, and tools — less if you build your own hive from scratch (find out how to do it in my book, *Building Beehives For Dummies,* published by Wiley). You'll spend around $100 or more for a package of bees and queen. For the most part, these are one-time expenses. Keep in mind, however, the potential for a return on this investment. Your hive can give you 60 to 90 pounds of honey every year. At around $8 per pound (a fair going price for all-natural, raw honey), that should give you an income of $480 to $720 per hive! Not bad, huh?

See Chapter 5 for a detailed listing of the equipment you'll need.

## How many hives do you need?

Most beekeepers start out with one hive. And that's probably a good way to start your first season. But most beekeepers wind up getting a second hive in short order. Why? For one, it's twice as much fun! Another more practical reason for having a second hive is that recognizing normal and abnormal situations is easier when you have two colonies to compare. In addition, a second hive enables you to borrow frames from a stronger, larger colony to supplement one that needs a little help. My advice? Start with one hive until you get the hang of things and then consider expanding in your second season.

## What kind of honey bees should you raise?

The honey bee most frequently raised by beekeepers in the United States today is European in origin and has the scientific name *Apis mellifera.*

Of this species, the most popular variety is the so-called Italian honey bee. These bees are docile, hearty, and good honey producers. They are a good choice for the new beekeeper. But there are notable others to consider. See Chapter 6 for more information about different varieties of honey bees.

## Time and commitment

Beekeeping isn't labor intensive. Sure, you'll spend part of a weekend putting together your new equipment. And I anticipate that you'll spend some time reading up on your new hobby. (I sure hope you read my book from cover to cover!) But the actual time that you absolutely *must* spend with your bees is surprisingly modest. Other than your first year (when I urge you to inspect the hive frequently to find out more about your bees), you need to make only five to eight visits to your hives every year. Add to that the time you spend harvesting honey, repairing equipment, and putting things away for the season, and you'll probably devote 35 to 40 hours a year to your hobby (more if you make a business out of it).

For a more detailed listing of seasonal activities, be sure to read Chapter 9.

## Beekeeper personality traits

If you scream like a banshee every time you see an insect, I suspect that beekeeping will be an uphill challenge for you. But if you love animals, nature, and the outdoors, and if you're curious about how creatures communicate and contribute to our environment, you'll be captivated by honey bees. If you like the idea of "farming" on a small scale, or you're intrigued by the prospect of harvesting your own all-natural honey, you'll enjoy becoming a beekeeper. Sure, as far as hobbies go, it's a little unusual, but all that's part of its allure. Express your uniqueness and join the ranks of some of the most delightful and interesting people I've ever met . . . backyard beekeepers!

## Allergies

If you're going to become a beekeeper, you can expect to get stung once in a while. It's a fact of life. But when you adopt good habits as a beekeeper, you can minimize or even eliminate the chances that you'll be stung.

All bee stings can hurt a little, but not for long. It's natural to experience some swelling, itching, and redness. These are *normal* (not allergic) reactions. Some folks are mildly allergic to bee stings, and the swelling and discomfort may be more severe. And yet, the most severe and life-threatening reactions to bee stings occur in less than 1 percent of the population. So the chances that you're dangerously allergic to honey bee venom are remote. If you're uncertain, check with an allergist, who can determine whether you're among the relatively few who should steer clear of beekeeping.

You'll find more information on bee stings in Chapter 3.

# Deciding Which Beekeeping Approach to Follow

Historically, beekeeping books have provided information on when and how to medicate your bee colonies and which chemicals to use for controlling pests that can compromise the health and productivity of your colonies. If you go online and visit beekeeping supply vendors, they all offer medications and pest control products to help bees when things go wrong.

But in recent years, because of all the problems that bees have been facing, it has become prudent to take a fresh look at these historical approaches to caring for and medicating your bees. Are treatments being overused? Probably. Are less-experienced beekeepers simply misusing these products to the detriment of our bees? Likely. Should you routinely medicate bees as so many traditional beekeeping books recommend? Doubtful. Or should you embrace a more *natural* approach with little or no use of medications or chemicals?

Clearly, there are many choices out there. This is a hot topic, and you will hear passionate arguments for and against the various options for various treatments. To decide which is right for you, it's first helpful to define each of the new approaches that are being discussed in today's world of beekeeping.

As a new beekeeper, you need to decide which approach or combination of approaches makes the most sense to you. I do my best to explain the options available to you and tell you about my own experiences. But it's up to you in the end. Pick an approach and stick to it until you find a better one. And be aware that if you ask ten beekeepers which is best, you will likely get ten different answers.

# Medicated beekeeping

*Medicated beekeeping* is a term intended to represent the "traditional" approach to honey-bee health that has been touted for decades in many of the books on beekeeping (including the first edition of this book). Indeed for generations, beekeepers were advised to follow a yearly protocol of medications and chemical treatments as part of the yearly routine. As bees faced more and more health issues, more and more chemical options came to market intended to help bees thrive. Many of these meds were administered prophylactically by well-intended beekeepers, just in case the bees might get ill — not because they needed it. I have no doubt that over time, with the growth of new hobbyist beekeepers, the increase in bee health problems, a plethora of new medications, and the possible overuse or misuse of these meds, the traditional ways outlined in so many earlier books will need to be rethought.

# Natural beekeeping

If you check the Internet, you'll find many opinions on what constitutes a *natural* approach to keeping bees. There is no universal definition. Natural beekeeping is more of an aspiration than an official set of rules. But still it's helpful to have a shorthand description that captures the *goal* of natural beekeeping. So I went to the expert. I asked Ross Conrad, author of *Natural Beekeeping, Revised and Expanded Edition* (Chelsea Green Publishing), to share his definition:

"When working on my book, my publisher and I settled on the title *Natural Beekeeping*. In retrospect, I realize that the term *natural beekeeping* is an oxymoron. A colony of bees that is manipulated by a person is no longer in its true, natural state. That said, the term natural beekeeping is used to refer to honey bee steward-ship that addresses pest, disease, and potential starvation issues without relying on synthetic pesticides, antibiotic drugs, or the regular use of an artificial diet."

Ross went on to tell me, "Natural beekeeping does not necessarily mean minimal manipulations and it definitely does not mean minimal hive inspections (as some have defined the term). If you are not regularly inspecting your colonies, you are unable to determine their needs, and you will be unable to take timely steps to keep your colonies viable. Minimal or no hive inspections is honey bee neglect, not natural bee stewardship."

## Organic beekeeping

Organic beekeeping is related to, but not the same as, natural beekeeping. There are a lot of written criteria available on what constitutes organic beekeeping. This material is currently being developed in minutiae for publication by various branches of the U.S. government and will ultimately be published as United States Department of Agriculture (USDA) standards to govern the production of organic honey and honey-related products. Under these guidelines, the use of some medications and chemical treatments is okay. To run a certified organic beekeeping operation, be prepared to take on a lot of work and make a sobering investment. Not too practical for the average backyard beekeeper. For the latest status of the new organic beekeeping regulations (known as Organic Apiculture Practice Standard, NOP-12-0063), visit the U.S. General Services Administration (GSA), Office of Information and Regulatory Affairs at www.reginfo.gov.

## Combining approaches

Okay. Here's my take on all of this. I don't personally follow any one of the medicated, natural, or organic approaches exclusively. In my view, there are no absolutes. I have no need to be certified as organic, so I choose not to go down that path. Generally speaking, I do not use chemicals "just in case" I may have a problem with pests. Nor do I typically medicate my bees as a preventive measure, but only when absolutely necessary, and only when other nonchemical options have not been effective. The same is true at home. I certainly don't take antibiotics whenever I feel sick or if I think I might get sick. But rest assured, if I came down with bacterial pneumonia, I would likely be asking my doc for antibiotics. And I certainly vaccinate my sweet golden retriever to keep her free of distemper. So my personal approach does not eliminate any use of medications, but rather follows a thoughtful, responsible approach that aspires to be as *natural* as possible. Like me, you may want to make choices based on what feels right to you.

**TIP**

In this edition, I have included lots of information that highlights alternative, more natural approaches to beekeeping than are found in books published in years past. Look for the All Natural icon to easily find my suggestions for those of you (like me) who are aspiring to minimize the use of medications and chemicals.

# Chapter **2**

# Getting to Know Your Honey Bees

My first introduction to life inside the honey bee hive occurred many years ago during a school assembly. I was about 10 years old. My classmates and I were shown a wonderful movie about the secret inner workings of the beehive. The film mesmerized me. I'd never seen anything so remarkable and fascinating. How could a bug be so smart and industrious? I couldn't help being captivated by the bountiful honey bee. That brief childhood event planted a seed that blossomed into a treasured hobby decades later.

Anyone who knows even a little bit about the honey bee can't help but be amazed, because far more goes on within the hive than most people can ever imagine: complex communication, social interactions, teamwork, unique jobs and respon-sibilities, food gathering, and the engineering of one of the most impressive living quarters found in nature. Whether you're a newcomer or an old hand, you'll have many opportunities to experience firsthand the miracle of beekeeping. Every time you visit your bees, you'll see something new. But you'll get far more out of your new hobby if you understand more about what you're looking at. What are the physical components of the bee that enable it to do its job so effectively? What are those bees up to and why? What's normal and what's not normal? What is a honey bee, and what is an imposter? In this chapter, you take a peek within a typical colony of honey bees.

# Basic Body Parts

Everyone knows about at least one part of the honey bee's anatomy: its stinger. But you'll get more out of beekeeping if you understand a little bit about the other various parts that make up the honey bee. I don't go into this in textbook-detail — I just show you a few basic parts to help you understand what makes honey bees tick. Figure 2-1 shows the basic parts of the worker bee.

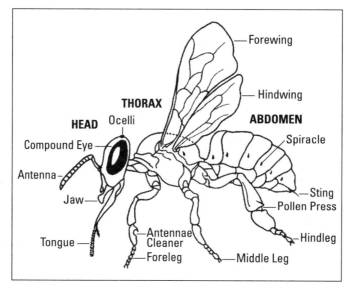

**FIGURE 2-1:**
This is how a honey bee looks if you shave off all the hairs. Several important body features are labeled.

*Courtesy of Howland Blackiston*

## Skeleton

Like all insects, the honey bee's skeleton is on the outside. This arrangement is called an *exoskeleton.* Nearly the entire bee is covered with *branched hairs* (like the needles on the branch of a spruce tree). Yes, the hairs help keep the bee body warm, but they also do much more. A bee can "feel" with these hairs, and the hairs serve the bee well when it comes to pollination because pollen sticks well to the branched hairs.

# Head

The honey bee's head is flat and somewhat triangular in shape. Here's where you find the bee's brain and primary sensory organs (sight, feel, taste, and smell). It's also where you find important glands that produce royal jelly and some of the various chemical pheromones used for communication. Figure 2-2 compares the heads of a worker, drone, and queen.

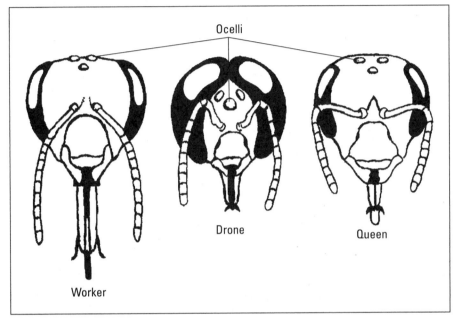

**FIGURE 2-2:** Comparing the heads of worker, drone, and queen bees. Note the worker bee's extra-long proboscis and the drone's huge, wraparound eyes.

*Royal jelly* is a substance secreted from glands in a worker bee's head and is used as a food to feed brood.

The important parts of the bee's head are its

>> **Eyes:** The head includes two large, compound eyes that are used for general-distance sight, and three small, simple eyes, called *ocelli,* which are used in the poor light conditions within the hive. Notice the three simple eyes (ocelli) on the members of all three castes in Figure 2-2, while the huge, wraparound, compound eyes of the drone make him easy to identify. The queen's eyes, however, are slightly smaller than the worker bee's.

>> **Antennae:** The honey bee has two antennae on the front of its face. Each antenna has thousands of tiny sensors that detect smell (like a nose does).

The bee uses this sense of smell to identify flowers, water, the colony, and maybe even you! The antennae also, like the branched hairs mentioned earlier, feel, detect moisture, measure distance when bees fly, and help the bee detect up and down, among other functions.

» **Mouth parts:** The bees' *mandibles* (jaws) are used for feeding larvae, collecting pollen, manipulating wax, and carrying things.

» **Proboscis:** Everyone's familiar with those noisemakers that show up at birthday and New Year's Eve parties. You know, the ones that unroll when you toot them! The bee's proboscis is much like those party favors, only without the "toot." When the bee is at rest, this organ is retracted and not visible. But when the bee is feeding or drinking, it unfolds to form a long tube that the bee uses like a straw.

## Thorax

The middle part of the bee is the *thorax*. It is the segment between the head and the abdomen where the two pairs of wings and six legs are anchored.

» **Wings:** Here's a question for you: How many wings does a honey bee have? The answer is four. Two pairs are attached fore and aft to the bee's thorax. The wings are hooked together in flight and separate when the bee is at rest.

» **Legs:** The bee's three pairs of legs are all different. Each leg has multiple segments that make the legs quite flexible. Bees also have taste receptors on the tips of their legs. The bee uses its forward-most legs to clean its antennae. The middle legs help with walking and are used to pack loads of pollen (and sometimes propolis) onto the *pollen baskets* that are part of the hind legs. *Propolis* is the sticky, resinous substance that the bees collect from the buds of trees and use to seal up cracks in the hive. Propolis can be harvested and used for a variety of nifty products. (For more information on propolis and what you can do with it, see Chapter 18.) The hind legs (see Figure 2-3) are specialized on the worker bee. They contain special combs and a pollen press, which are used by the worker bee to brush, collect, pack, and carry pollen and propolis back to the hive. Take a moment to watch a foraging bee on a flower. You'll see her hind legs heavily loaded with pollen for the return trip home.

» **Spiracles:** These tiny holes along the sides of a bee's thorax and abdomen are the means by which a bee breathes. The bee's trachea (breathing tubes) are attached to these spiracles. Tracheal mites gain access to the trachea through the first hole in the thorax. These mites are a problem for bees; you find out how to deal with tracheal mites in Chapter 13.

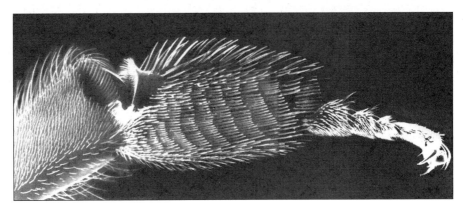

**FIGURE 2-3:**
In this close-up image of a bee's leg, you can clearly see the hairs that serve as brushes to collect pollen.

*Courtesy of Dr. Eric Erickson, Jr.*

## Abdomen

The abdomen is the part of the bee's body that contains its digestive organs, heart, reproductive organs, wax and scent glands (workers only), and, of course, the infamous stinger (workers and queen only).

# The Amazing Language of Bees

It is said that only man and primates have a form of communication superior to that of the honey bee. Like you and I, honey bees use five senses throughout their daily lives; however, honey bees have additional communication aids at their disposal. Two of the methods by which they communicate are of particular interest. One is chemical, the other choreographic.

## Pheromones

*Pheromones* are chemical scents that animals produce to trigger behavioral responses from the other members of the same species. Honey bee pheromones provide the "glue" that holds the colony together. The three types of adults in a hive, of which more is mentioned later in this chapter, produce different pheromones at different times to stimulate specific behaviors. The study of pheromones is a topic worthy of an entire book, so here are just a few basic facts about the ways pheromones help bees communicate:

>> Certain queen pheromones (known as *queen substance,* discussed at greater length later in this chapter) let the entire colony know that the queen is in residence and stimulate many worker-bee activities.

>> Outside of the hive, the queen pheromones act as a sex attractant to potential suitors (male drone bees from other colonies). They also help regulate the drone (male bee) population in the hive.

>> Queen pheromones stimulate many worker-bee activities, such as comb-building, brood-rearing, foraging, and food storage.

>> The worker bees at the hive's entrance produce pheromones that help guide foraging bees back to their hive. The Nasonov gland (discussed later in this chapter) at the tip of the worker bee's abdomen is responsible for this alluring scent.

>> Worker bees produce alarm pheromones that can trigger sudden and decisive aggression from the colony.

>> The colony's brood (developing bee larvae and pupae) secretes special pheromones that help worker bees recognize the brood's gender, stage of development, and feeding needs.

## Shall we dance?

Perhaps the most famous and fascinating "language" of the honey bee is communicated through a series of dances done by foraging worker bees who return to the hive with news of nectar, pollen, or water. The worker bees dance on the comb using precise patterns. Depending on the style of dance, a variety of information is shared with the honey bees' sisters. They're able to obtain remarkably accurate information about the location and type of food the foraging bees have discovered.

Two common types of dances are the so-called *round dance* and the *waggle dance.*

The round dance communicates that the food source is near the hive (within 10–80 yards). Figure 2-4 illustrates dancing movements.

For a food source found at a greater distance from the hive, the worker bee performs the waggle dance. It involves a shivering side-to-side motion of the abdomen while the dancing bee forms a figure eight. The vigor of the waggle, the number of times it is repeated, the direction of the dance, and the sound the bee makes communicate amazingly precise information about the location of the food source. See Figure 2-4.

The dancing bees pause between performances to offer potential recruits a taste of the goodies they bring back to the hive. Combined with the dancing, the samples provide additional information about where the food can be found and what type of flower it is from.

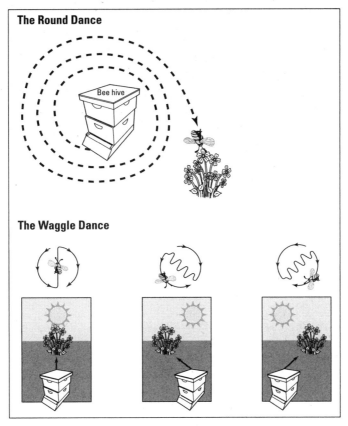

**FIGURE 2-4:** The round dance (top) and the waggle dance (bottom).

*Courtesy of Howland Blackiston*

# Getting to Know the Male and the Two Female Castes

During summer months, about 60,000 or more bees reside in a healthy hive. And while you may think that all those insects look exactly alike, the population actually includes two different female castes (the queen and the workers) and the male bees (drones; see Figure 2-5). Each type has its own characteristics, roles, and responsibilities. Upon closer examination, the three even look a little different, and being able to distinguish one from the other is important.

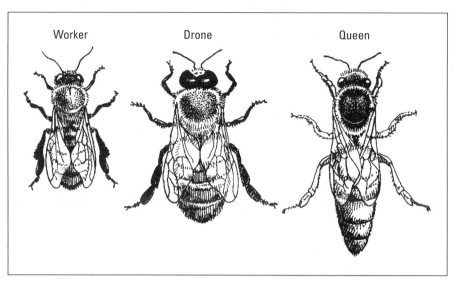

Worker    Drone    Queen

*Courtesy of Howland Blackiston*

**FIGURE 2-5:**
These are the three types of bees in the hive: worker, drone, and queen.

## Her majesty, the queen

Let there be no mistake about it — the queen bee is the heart and soul of the colony. She is the reason for nearly everything the rest of the colony does. The queen is the only bee without which the rest of the colony cannot survive. Without her, your hive is sunk. A good-quality queen means a strong and productive hive. For more information on how to evaluate a good queen, see Chapter 8. And for some real fun, try raising your own queens from your best performing hives. See Chapter 14.

**REMEMBER**

As a beekeeper, on every visit to the hive you need to determine two things: "Do I have a queen?" and "Is she healthy?"

Only one queen lives in a given hive. She is the largest bee in the colony, with a long and graceful body. She is the only female with fully developed ovaries. The queen's two primary purposes are to produce chemical scents that help regulate the unity of the colony and to lay eggs — and lots of them. She is, in fact, an egg-laying machine, capable of producing more than 1,500 eggs a day at 30-second intervals. That many eggs are more than her body weight!

The other bees pay close attention to the queen, tending to her every need. Like a regal celebrity, she's always surrounded by a flock of attendants (her *retinue*) as she moves about the hive (see Figure 2-6). Yet, she isn't spoiled. These attendants are vital because the queen is totally incapable of tending to her own basic needs.

She can neither feed nor groom herself. She can't even leave the hive to relieve herself. And so her doting attendants (the queen's court) take care of her basic needs while she tirelessly goes from cell to cell doing what she does best . . . laying eggs.

Courtesy of USDA-ARS

FIGURE 2-6:
A queen and her attentive attendants.

The gentle queen bee has a stinger, but it is rare for a beekeeper to be stung by a queen bee. I have handled many queen bees and have never been stung by any of them. In general, queen bees use their stingers only to kill rival queens that may emerge or be introduced in the hive.

The queen can live for two or more years, but replacing your queen after a season or two ensures maximum productivity and colony health. Many seasoned beekeepers routinely replace their queens every year after the nectar flow. This practice ensures that the colony has a new, energetic, and fertile young queen each season. You may wonder why you should replace the queen if she's still alive. That's an easy one: As a queen ages, her egg-laying capability slows down, which results in less and less brood each season. Less brood means a smaller colony. And a smaller colony means a lackluster honey harvest for you! For information on how to successfully introduce a new queen, see Chapter 10. For information on how to raise your own queens (now, that's fun!), see Chapter 14.

REMEMBER

As a beekeeper, your job is to *anticipate* problems before they happen. An aging queen — more than a year old — is something that you can deal with by replacing her after checking her egg-laying, before you have a problem.

# AMAZING "QUEEN SUBSTANCES"

In addition to laying eggs, the queen plays a vital role in maintaining the colony's cohesiveness and stability. The mere presence of the queen in the hive motivates the productivity of the colony. Her importance to the hive is evident in the amount of attention paid to her by the worker bees everywhere she goes in the hive. But, as is true of every working mom or regal presence, she can't be everywhere at once, and she doesn't interact with every member of the colony every day. So how does the colony know it has a queen? By her scent. The queen produces a number of different pheromones (mentioned earlier in this chapter) that attract workers to her and stimulate brood-rearing, foraging, comb-building, and other activities. Also referred to as *queen substances,* these pheromones play an important role in controlling the behavior of the colony: Queen substances keep the worker bees from making a new queen and inhibit the development of the worker bees' ovaries, thus ensuring that the queen is the only egg-laying female in the hive. They act as a chemical communication that "all is well — the queen is in residence and hard at work." As a queen ages, these pheromones diminish, and when that happens, the colony knows it's time to supersede her with a new, young queen.

Pheromones are essential in controlling the colony's well-being. This queen substance makes its way around the hive like a bucket brigade. The queen's attendants pick up the scent from the queen and transfer it by contact with neighboring bees. They in turn pass the scent onto others, and so it distributes throughout the colony. So effective is this relay that if the queen were removed from the hive, the entire colony would be aware of her loss within hours. When the workers sense the lack of a queen, they become listless, and their drive to be productive is lost. Without leadership, they nearly lose their reason for being! First they're unhappy and mope around, but then it dawns on them . . . "Let's make a new queen."

## The industrious little worker bee

The majority of the hive's population consists of *worker bees.* Like the queen, worker bees are all female. Worker bees that are younger than 3 weeks old have working ovaries and can lay eggs, but they are not fertile, as the workers never mate and store no sperm. Workers also look different than the queen. They are smaller, their abdomens are shorter, and on their hind legs they possess *pollen baskets,* which are used to tote pollen back from the field.

Like the queen, the worker bee has a stinger. But her stinger is not a smooth syringe like the queen's. The sting is three-shafted, with each shaft having a barb (like a fish hook). The barbs cause the stinger, venom sack, and a large part of the bee's gut to remain in a human victim — a Kamikaze effort to protect the colony. Only in mammals (such as humans) does the bee's stinger get stuck. The worker bee can sting other insects again and again while defending its home.

The life span of a worker bee is a modest six weeks during the colony's active season. However, worker bees live longer (four to eight months) during the less-active winter months. These winter workers are loaded with protein and are sometimes referred to as "Fat Bees." The term "busy as a bee" is well earned. Worker bees do a considerable amount of work, day in and day out. They work as a team. Life in the hive is one of compulsory cooperation. What one worker could never do on her own can be accomplished as a colony. During the busy season, the worker bees literally work themselves to death. The specific jobs and duties they perform during their short lives vary as they age. Understanding their roles will deepen your fascination and appreciation of these remarkable creatures.

From the moment a worker bee emerges from her cell, she has many and varied tasks clearly cut out for her. As she ages, she performs more and more complex and demanding tasks. Although these various duties usually follow a set pattern and timeline, they sometimes overlap. A worker bee may change occupations, sometimes within minutes, if there is an urgent need within the colony for a particular task. They represent teamwork and empowerment at their best!

Initially, a worker's responsibilities include various tasks within the hive. At this stage of development, worker bees are referred to as *house bees.* As they get older, their duties involve work outside of the hive as *field bees.*

# House bees

Worker honey bees spend the first few weeks of their lives carrying out very specific tasks *within* the hive. For this reason, they are referred to as *house bees.* The jobs they do (described in the following sections) are dependent on their age.

## Housekeeping (days 1 to 3)

A worker bee is born with the munchies. Immediately after she emerges from the cell and grooms herself, she engorges herself with pollen and honey. Following this binge, one of her first tasks is cleaning out the cell from which she just emerged. This cell and other empty cells are cleaned and polished and left immaculate to receive new eggs and to store nectar and pollen.

## Undertaking (days 3 to 16)

The honey bee hive is one of the cleanest and most sterile environments found in nature. Preventing disease is an important early task for the worker bee. During the first couple weeks of her life, the worker bee removes any bees that have died and disposes of the corpses as far from the hive as possible. Similarly, diseased or dead brood are quickly removed before becoming a health threat to the colony.

Should a larger invader (such as a mouse) be stung to death within the hive, the workers have an effective way of dealing with that situation. Obviously a dead mouse is too big for the bees to carry off. So the workers completely encase the corpse with *propolis* (a brown, sticky resin collected from trees and sometimes referred to as *bee glue*). Propolis has significant antibacterial qualities. In the hot, dry air of the hive, the hermetically sealed corpse becomes mummified and is no longer a source of infection. The bees also use propolis to seal cracks and varnish the inside walls of the hive.

## Working in the nursery (days 4 to 12)

The young worker bees tend to their baby sisters by feeding and caring for the developing larvae. On average, nurse bees check a single larva 1,300 times a day. They feed the larvae a mixture of pollen and honey, and royal jelly — rich in protein and vitamins — produced from the hypopharyngeal gland in the worker bee's head. The number of days spent tending brood depends on the quantity of brood in the hive and the urgency of other competing tasks.

## Attending royalty (days 7 to 12)

Because her royal highness is unable to tend to her most basic needs by herself, some of the workers do these tasks for her. They groom and feed the queen and even remove her excrement from the hive. These royal attendants also coax the queen to continue to lay eggs as she moves about the hive.

## Going grocery shopping (days 12 to 18)

Young worker bees also take nectar from foraging field bees that are returning to the hive. The house bees deposit this nectar into cells earmarked for this purpose. They add an enzyme to the nectar and set about fanning the cells to evaporate the water content and turn the nectar into ripened honey. The workers similarly take pollen from returning field bees and pack the pollen into other cells. Both the ripened honey and the pollen, which is often referred to as *bee bread,* are food for the colony.

## Fanning (days 12 to 18)

Worker bees also take a turn at controlling the temperature and humidity of the hive. During warm weather and during the honey flow season, you'll see groups of bees lined up at one side of the entrance, facing the hive. They fan furiously to draw air into the hive. Additional fanners are in position within the hives. This relay of fresh air helps maintain a constant temperature (93 to 95 degrees Fahrenheit [34 to 35 degrees Celsius]) for developing brood. The fanning also hastens the evaporation of excess moisture from the curing honey. Chapter 15 takes a detailed look at what makes honey so special.

# ROYAL JELLY: THE FOOD OF ROYALTY

*Royal jelly* is the powerful, creamy substance that transforms an ordinary worker-bee egg into a queen bee and extends her life span from six weeks to several years! It's made of digested pollen and honey or nectar mixed with a chemical secreted from a gland in a nurse bee's head.

In health-food stores it commands premium prices, rivaling imported caviar. Those in the know use royal jelly as a dietary supplement and fertility stimulant. It contains an abundance of nutrients, including essential minerals, B-complex vitamins, proteins, amino acids, collagen, and essential fatty acids, just to name a few!

The workers also perform another kind of fanning, but it isn't related to climate control. It has more to do with communication. The bees have a scent gland located at the end of their abdomen called the *Nasonov gland.* You'll see worker bees at the entrance with their abdomens arched and the moist pink membrane of this gland exposed (see Figure 2-7). They fan their wings to release this pleasant, sweet odor into the air. You can actually smell it sometimes as you approach the hive. The pheromone is highly attractive and stimulating to other bees and serves as an orientation message to returning foragers, saying: "Come hither . . . this is your hive and where you belong."

**FIGURE 2-7:** This worker bee fans her wings while exposing her Nasonov gland to release a sweet orientation scent. This helps direct other members of the colony back to the hive.

*Courtesy of Bee Culture Magazine*

Beekeepers can purchase synthetic queen-bee pheromone and use this chemical to lure swarms of bees into a trap. The captured swarm then can be used to populate a new hive.

### Becoming architects and master builders (days 12 to 35)

Worker bees that are about 12 days old are mature enough to begin producing beeswax. These white flakes of wax are secreted from wax glands on the underside of the worker bee's abdomen. They help with the building of new *wax comb* and in the capping of ripened honey and cells containing developing pupae.

Some new beekeepers are alarmed when they first see these wax flakes on the bee. They wrongly think these white chips are an indication of a disease or mite problem.

### Guarding the home (days 18 to 21)

The last task of a house bee before she ventures out is that of guarding the hive. At this stage of maturity, her sting glands have developed to contain an authoritative amount of venom. You can easily spot the guard bees at the hive's entrance. They are poised and alert, checking each bee that returns to the hive for a familiar scent. Only family members are allowed to pass. Strange bees, wasps, hornets, and others intent on robbing the hive's vast stores of honey are bravely driven off.

Bees from other hives are occasionally allowed in when they bribe the guards with nectar. These bees simply steal a little honey or pollen and then leave.

## Field bees

When the worker bee is a few weeks old, she ventures outside the hive to perform her last and perhaps most important job — to collect the pollen and nectar that will sustain the colony. With her life half over, she joins the ranks of field bees until she reaches the end of her life.

It's not unusual to see field bees taking their first *orientation flights.* The bees face the hive and dart up, down, and all around the entrance. They're imprinting the look and location of their home before beginning to circle the hive and progressively widening those circles, learning landmarks that ultimately will guide them back home. At this point, worker bees are foraging for pollen (see Figure 2-8), nectar, water, and propolis (resin collected from trees).

**FIGURE 2-8:**
This bee's pollen baskets are filled. She can visit ten flowers every minute and may visit more than 600 flowers before returning to the hive.

Foraging bees visit 5 million flowers to produce a single pint of honey. They forage a 1- to 2-mile radius from the hive in search of food (even more if necessary). That's the equivalent of several thousand acres! So don't think for a moment that you need to provide everything they need on your property. They're ready and willing to travel.

Foraging is the toughest time for the worker bee. It's difficult and dangerous work, and it takes its toll. They can get chilled as dusk approaches and die before they can return to the hive. Sometimes they become a tasty meal for a bird or other insect. You can spot the old girls returning to the hive. They've grown darker in color, and their wings are torn and tattered. This is how the worker bee's life draws to a close . . . working diligently right until the end.

## The woeful drone

This brings us to the drone, the male bee in the colony. Drones make up a relatively small percentage of the hive's total population. At the peak of the season, their numbers may be only in the hundreds. You rarely find more than a thousand.

New beekeepers often mistake a drone for the queen, because he is larger and stouter than a worker bee. But his shape is in fact more like a barrel (the queen's shape is thinner, more delicate, and tapered). The drone's eyes are huge and seem to cover his entire head. He doesn't forage for food from flowers — he has no pollen baskets. He doesn't help with the building of comb — he has no wax-producing glands. Nor can he help defend the hive — he has no stinger. He is not the queen or a worker — merely the drone.

The drone gets a bad rap in many bee books. Described as lazy, glutinous, and incapable of caring for himself, you might even begin wondering what he's good for.

He mates! Procreation is the drone's primary purpose in life. Despite their high maintenance (they must be fed and cared for by the worker bees), drones are tolerated and allowed to remain in the hive because they may be needed to mate with a new virgin queen (when the old queen dies or needs to be superseded). Mating occurs outside of the hive in mid-flight, 200 to 300 feet in the air. This location is known as the "Drone Congregation Area," and it can be a mile or more away from the hive. The drone's big eyes come in handy for spotting virgin queens (from other colonies) who are taking their *nuptial flights.* The few drones that do get a chance to mate are in for a sobering surprise. They die after mating! That's because their sex organ is barbed (like the worker bee's stinger) so they can discharge their sperm into the queen. An organ inside the queen called the *spermatheca* is the receptacle for the sperm. The queen will mate with several drones during her nuptial flight. After mating with the queen, the drone's most personal apparatus and a significant part of its internal anatomy is torn away, and it falls to its death, a fact that prompts empathetic groans from the men in my lectures and some unsympathetic cheers from a few women.

Once the weather gets cooler and the mating season comes to a close, the workers do not tolerate having drones around. After all, those fellows have big appetites and would consume a tremendous amount of food during the perilous winter months. So in cooler climates at the end of the nectar-producing season, the worker bees systematically expel the drones from the hive (see the photo in this book's color section). Drones are literally tossed out the door. For those beekeepers who live in areas that experience cold winters, this is your signal that the beekeeping season is over for the year.

Depending on where you live, the calendar of events for you and your bees varies depending on temperature ranges and the time of year. To read more about the beekeeper's calendar in your part of the world, see the information and chart in Chapter 9.

# The Honey Bee Life Cycle

In regions where winter means "cold," the hive is virtually dormant. The adult bees are in a tight cluster for warmth, and their queen is snugly safe in the center of it all. But as the days lengthen and the spring season approaches, the bees begin feeding the queen royal jelly. This special food (secreted from the glands near the workers' mandibles) is rich in protein and stimulates the queen to start laying eggs.

Like butterflies, honey bees develop in four distinct phases: egg, larva, pupa, and adult. The total development time varies a bit among the three types of bees, but the basic miraculous process is the same: 16 days for queens, 21 days for worker bees, and 24 days for drones.

## Egg

The metamorphosis begins when the queen lays an egg. You should know how to spot eggs because that is one of the most basic and important skills you need to develop as a beekeeper. It isn't an easy task because the eggs are mighty tiny (only about 1.7 millimeters long). But finding eggs is one of the surest ways to confirm that your queen is alive and well. It's a skill you'll use just about every time you visit your hive.

The queen lays a single egg in each cell that has been cleaned and prepared by the workers to raise new brood (see Figure 2-9). The cell must be spotless, or she moves on to another one.

If she chooses a standard worker-size cell, she releases a fertilized egg into the cell. That egg develops into a worker bee (female). But if she chooses a wider, drone-size cell, the queen releases a nonfertilized egg. That egg develops into a drone bee (male). The workers that build the cells regulate the ratio of female worker bees to male drone bees. They do this by building smaller cells for female worker bees and larger cells for male drone bees.

Having said all that, not all fertilized eggs develop into worker bees. Some can develop into a regal queen bee. But more on that in Chapter 14.

The queen positions the egg in an upright position (standing on end) at the bottom of a cell. That's why they're so hard to see. When you look straight down into the cell, you're looking at the minuscule diameter of the egg, which is only 0.4 millimeter wide. Figure 2-10 shows a microscopic close-up of a single egg.

**FIGURE 2-9:**
Note the rice-like shape of the eggs and how the queen has positioned them standing up in the cells.

*Courtesy of Howland Blackiston*

**FIGURE 2-10:**
A single egg.

*Courtesy of Alexander Wild, www.alexanderwild.com*

TIP

Eggs are much easier to spot on a bright, sunny day. Hold the comb at a slight angle, with the sun behind you and shining over your shoulder, illuminating the deep recesses of the cell. The eggs are translucent white and resemble a miniature grain of rice. I recommend that you invest in an inexpensive pair of reading glasses. The magnification can really help you spot the eggs (even if you don't normally need reading glasses). After you discover your first egg, it'll be far easier to know what you're looking for during future inspections. Better yet, get yourself a pair of magnifying goggles such as those used by watchmakers and model makers (see Figure 2-11).

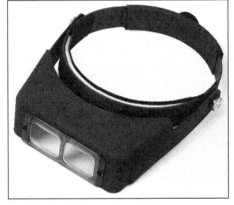

**FIGURE 2-11:**
These magnifying goggles (used by watchmakers and model makers) are a great beekeeping tool for finding those itty-bitty eggs.

## Larva

Three days after the queen lays the egg, it hatches into a *larva* (the plural is *larvae*). Healthy larvae are glistening and snowy white and resemble small grubs curled up in the cells (see Figure 2-12). Tiny at first, the larvae grow quickly, shedding their skin five times. These helpless little creatures have voracious appetites, consuming meals 24 hours a day. The nurse bees first feed the larvae royal jelly, and later they're weaned to a mixture of honey and pollen (sometimes referred to as *bee bread*). Within just five days, they are 1,500 times larger than their original size. At this time the worker bees seal the larvae in the cell with a porous capping of tan beeswax. Once sealed in, the larvae spin a cocoon around their bodies.

## Pupa

The larva is now officially a *pupa* (the plural is *pupae*). Here's where things really begin to happen. Of course the transformations now taking place are hidden from sight under the wax cappings. But if you could, you'd see that this little creature is beginning to take on the familiar features of an adult bee (see Figure 2-13).

The eyes, legs, and wings take shape. Coloration begins with the eyes: first pink, then purple, and then black. Finally, the fine hairs that cover the bee's body develop. After 12 days (in the case of the worker) the now *adult bee* chews her way through the wax capping to join her sisters and brothers. Figure 2-14 shows the entire life cycle of the three castes of honey bee from start to finish.

**FIGURE 2-12:** Beautiful, pearly white little larvae curled up in their cells.

*Courtesy of John Clayton*

**FIGURE 2-13:** Opened cells reveal an egg and developing pupae.

*Courtesy of Dr. Edward Ross, California Academy of Sciences*

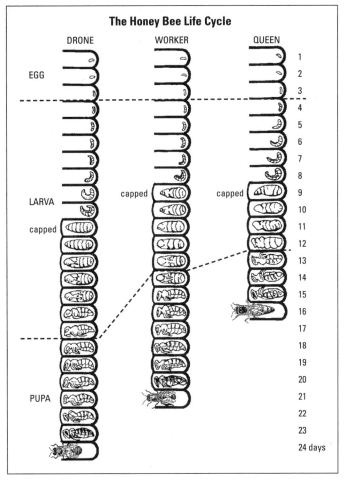

**FIGURE 2-14:**
This chart shows the daily development cycle of the two female castes and the drone, from egg to adult.

*Courtesy of Howland Blackiston*

# Other Stinging Insects

Many people are quick to say they've been "stung by a bee," but the chances of a *honey* bee stinging them are rather slim. Honey bees usually are gentle in nature, and it is rare for an individual to be randomly stung by a honey bee. Honey bees sting as a defensive behavior; they are most likely to sting around the hive or when disturbed while foraging on a flower.

Other stinging insects, especially some of the wasps, are the more likely culprits when someone is stung. Most folks, however, don't make the distinction between honey bees and wasps. They incorrectly lump all insects with stingers into the "bee" category. True bees are unique in that their bodies are covered with hair, and they use pollen and nectar from plants as their sole source of food (they're not the ones raiding your cola drink at a picnic — those are likely to be yellow-jacket wasps). Here are some of the most common stinging insects.

## Bumblebee

The gentle bumblebee, sometimes called a *humble* bee (see Figure 2-15), is large, plump, mostly black and yellow, and hairy. It's a familiar sight, buzzing loudly from flower to flower, collecting pollen and nectar. Bumblebees live in small ground nests, and all but the queen die off every autumn. At the peak of summer, the colony is only a few hundred strong. Bumblebees make honey, but only small amounts (measured in ounces, not pounds). They are generally quite docile and not inclined to sting unless they or their nest is disturbed.

**FIGURE 2-15:**
The bumblebee is furry and plump.

*Courtesy of Dr. Edward Ross, California Academy of Sciences*

## Carpenter bee

The carpenter bee (see Figure 2-16) looks much like a bumblebee, but its habits are quite different. It is a solitary bee that makes its nest by tunneling through solid wood (sometimes the wooden eaves of a barn or shed). Like the honey bee, the carpenter bee forages for pollen. Its nest is small and produces only a few dozen offspring a season. Carpenter bees are gentle and are not likely to sting unprovoked. But they can do some serious damage to the woodwork on your house.

**FIGURE 2-16:**
The carpenter
bee looks similar
to a bumblebee,
but its abdomen
has no hair.

*Courtesy of Dr. Edward Ross, California Academy of Sciences*

## Mason bee

A bit similar in appearance to honey bees, the mason bee is a very important contributor to effective pollination. Unlike honey bees or bumblebees, *mason bees* are solitary, meaning there is no queen bee, and every female is fertile, making her own nest. Mason bees produce neither honey nor beeswax. In the wild, they build their mud-sealed nests in natural tubes like reeds or holes in dead trees. There has been a considerable insurgence of interest in these lovely insects, and commercially produced nests are available to house mason bees for the pollination of gardens and orchards (see Figure 2-17). Although mason bees do have stingers, they are very gentle. Even if you do get stung, it is more like a mosquito bite than a painful bee, wasp, or hornet sting.

## Wasp

Many different kinds of insects are called *wasps.* The more familiar of these are distinguished by their smooth, hard bodies (usually brown or black) and familiar ultra-thin wasp waist (see Figure 2-18). So-called social wasps build exposed paper nests, which usually are rather small and contain only a handful of insects and brood, from wood fiber. These nests sometimes are located where we'd rather not have them (in a door frame or windowsill). The slightest disturbance can lead to defensive behavior and stings. Social wasps primarily are meat eaters, but adult wasps are attracted to sweets. Note that wasps and hornets have smooth stingers (no barbs) and can inflict their fury over and over again. Ouch!

**FIGURE 2-17:**
A commercially
available
mason-bee nest.

*Courtesy of Brushy Mountain Bee Farm*

**FIGURE 2-18:**
The wasp is
clearly identified
by its smooth,
hairless body and
narrow wasp
waist.

*Courtesy of Dr. Edward Ross, California Academy of Sciences*

## Yellow jacket

The yellow jacket also is a social wasp. Fierce and highly aggressive, it is likely responsible for most of the stings wrongly attributed to honey bees (see Figure 2-19). Yellow jackets are a familiar sight at summer picnics, where they scavenge for food and sugary drinks. Two basic kinds of yellow jackets exist: those that build their paper nests underground (which can create a problem when noisy lawn mowers or thundering feet pass overhead) and those that make their nests in trees. All in all, yellow jackets aren't very friendly bugs.

**FIGURE 2-19:** The ill-tempered yellow jacket is a meat eater but also has a taste for sweets.

*Courtesy of Dr. Edward Ross, California Academy of Sciences*

## Bald-faced hornet

Bald-faced hornets are not loveable creatures (see Figure 2-20). Nor are they true hornets. Typically found in the eastern United States, they are in fact aerial-nesting yellow jackets, building their nests above ground (in trees or bushes). They are highly defensive, have a painful sting, and are ruthless hunters and meat eaters. They do, however, build fantastically impressive and beautiful paper nests from their saliva and wood fiber they harvest from dead trees (see Figure 2-21). These nests can grow large during the summer and eventually reach the size of a basketball. Such nests can contain several thousand large, hot-tempered hornets — keep your distance! The end of the summer marks the end of the hornet city. When the cool weather approaches, the nest is abandoned, and only a newly raised queen survives. She finds a warm retreat underground and emerges in the spring, raising young and building a new nest.

**FIGURE 2-20:**
The bald-faced hornet makes impressive paper nests in trees.

*Courtesy of Dr. Edward Ross, California Academy of Sciences*

**FIGURE 2-21:**
A large paper nest made by a colony of bald-faced hornets.

*Courtesy of Howland Blackiston*

# 2

# Starting Your Adventure

Overcome any initial fears and apprehensions of becoming a beekeeper.

Discover how to conquer potential roadblocks from local authorities, family members, and your immediate neighbors.

Understand the basics of where to locate your hives, when to start your adventure, and how to feed and water your bees.

Find out about six different and popular types of hives, and understand the strengths and weaknesses of each design.

Become familiar with the various tools and equipment that you will be using as a new backyard beekeeper.

Determine the breed of honey bee that will work best for you and find out where to obtain your bees, when to order them, and how to safely install them into your new hive.

IN THIS CHAPTER

» Avoiding the dreaded stinger

» Understanding local restrictions

» Winning over your family, friends, neighbors, and landlords

» Picking the perfect location

» Deciding the best time to start

# Chapter **3**

# Alleviating Apprehensions and Making Decisions

I suspect all new backyard beekeepers think similar thoughts as they're deciding to make the plunge. You've thought about beekeeping for some time. You're growing more and more intrigued by the idea . . . maybe this is the year you're going to do something. It certainly sounds like a lot of fun. What could be more unique? It's educational and a nice outdoor activity for you — back to nature and all that stuff. The bees will do a great job of pollinating the garden, and there's that glorious crop of delicious homegrown honey to look forward to. And you realize you can make a difference by introducing a colony of bees in a time when the feral bee population is in jeopardy. The anticipation is building daily, and you're consumed with excitement. That's it! You've made up your mind. You'll become a beekeeper! But in the back of your mind, some nagging concerns keep bubbling to the surface.

You're a wee bit concerned about getting stung, aren't you? Your friends and family may say you're crazy for thinking of becoming a beekeeper. What if the neighbors disapprove when they find out? Maybe bees are not even *allowed* in your neighborhood. What happens if the bees don't like their new home and all fly away? Help!

Relax. These are certainly some of the concerns I had when I first started. In this chapter, I hope to defuse your apprehensions and suggest some helpful ways to deal with those concerns.

# Overcoming Sting Phobia

Perhaps the best-known part of the bee's anatomy is its stinger. Quite honestly, that was my biggest apprehension about taking up beekeeping. I don't think I'd ever been stung by a honey bee, but I'd certainly felt the wrath of yellow jackets and wasps. I wanted no part of becoming a daily target for anything so unpleasant. I fretted about my fear for a long time, looking for reassurances from experienced beekeepers. They told me time and again that honey bees bred for beekeeping were docile and seldom inclined to sting. But lacking firsthand experience, I was doubtful.

The advice turned out to be 100 percent correct. Honey bees *are* docile and gentle creatures. To my surprise (and delight), I made it through my entire first season without receiving a single sting. In the decades that I've been keeping bees, not a single member of my family, not a single visitor to my home, and not a single neighbor has ever been stung by one of my honey bees.

By the way, bees *sting* — they don't *bite*. Honey bees use their stinger only as a last resort to defend the colony. After all, they die after stinging a person. When bees are away from the hive (while they're collecting nectar and pollen), defending the colony is no longer a priority, so they're as gentle as lambs out in the field, unless, of course, accidently stepped on.

Do I ever get stung? Sure. But usually not more than three or four times a year. In every case, the stings I take are a result of my own carelessness. I'm rushing, taking shortcuts, or am inattentive to their mood — all things I shouldn't do. (I really should read this book more often!) My sloppiness is merely the result of becoming so comfortable with my bees that I'm not as diligent as I should be. The secrets to avoiding stings are your technique and demeanor.

Here are some helpful tips for avoiding stings:

>> Always wear a veil and use your smoker when visiting your hive (see Chapter 5 for more information on these two vital pieces of beekeeper apparatus).

>> Inspect your bees during pleasant daytime weather. Try to use the hours between 10 a.m. and 5 p.m. That's when most of the bees are out working, and fewer bees are at home. Don't open up the hive at night, during bad

weather, or if a thunderstorm is brewing. In Chapters 7 and 8, I go into detail about how to open the hive and inspect the colony.

Keep in mind that on city rooftops, the wind chill can lower the air temperature considerably. Be aware of its damaging effect on a developing brood.

**URBAN**

>> Don't rush. Take your time and move calmly. Sudden movements are a no-no.

>> Get a good grip on frames. If you drop a frame of bees, you'll have a memorable story to tell.

>> Never swat at bees. Become accustomed to them crawling on your hands and clothing. They're just exploring. Bees can be gently pushed aside if necessary.

>> When woodenware is stuck together with propolis, don't snap it apart with a loud "crack." The bees go on full alert when they feel sudden vibrations.

**URBAN**

Bees don't like loud noises and vibrations, so don't locate your city bees near noisy compressors, air conditioners, or anything else that might make them cranky.

>> Never leave sugar syrup or honey in open containers near the hive. Doing so can excite bees into a frenzy, and you may find yourself in the middle of it. It can also set off *robbing* — an unwelcome situation in which bees from other colonies attack your bees, robbing them of their honey. In Chapter 10, you find instructions on how to avoid robbing and what to do when it happens.

>> Keep yourself and your bee clothing clean. Bees don't like bad body odor. If you like to eat garlic, avoid indulging right before visiting your bees. Chapter 7 has some handy hygiene hints.

>> Wear light-colored clothing. Bees don't seem to like dark colors.

## Knowing what to do if you're stung

Be prepared to answer the following question from everyone who hears you're a beekeeper: "Do you ever get stung?" You'll hear this a hundred times. An occasional sting is a fact of life for a beekeeper. Following the rules of the road, however, keeps stings to a minimum, or perhaps you'll get none at all. Yet, if a bee stings you or your clothing, calmly remove the stinger and smoke the area to mask the chemical alarm scent left behind. (This alarm pheromone can stimulate other bees to sting.) To remove the stinger, you can use your fingernail to *scrape* it off your skin.

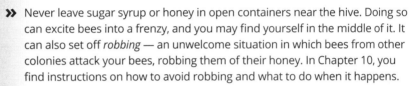

**TIP**

If you are stung, apply a cold compress and take an antihistamine tablet (such as Benadryl). Antihistamine creams also are available. Using this technique reduces the swelling, itching, and discomfort.

Some folks swear by the effectiveness of baking-soda-and-water poultices for bee stings; other folks advocate meat tenderizer and wet tobacco poultices, respectively. These are "grandma recipes" that were used before we had the antidote that the medical profession endorses — over-the-counter antihistamines.

## Watching for allergic reactions

All bee stings hurt a bit, but not for long. Experiencing redness, swelling, and itching is completely natural. These are normal (not allergic) reactions. For a small percentage of individuals, more severe allergic or even toxic reactions can occur, including severe swelling beyond the immediate area of the sting and shortness of breath. In the worst cases, reaction to bee stings can result in loss of consciousness or even death. The most severe reactions occur in less than 1 percent of the population. To put that in perspective, more people are killed by lightning each year than die from bee stings.

As a precaution against a guest having a severe reaction, I keep an EpiPen on hand. These emergency sting remedies are available from your doctor by prescription. The EpiPen automatically injects a dosage of epinephrine (adrenaline). But be careful. Liability issues can arise when injecting another person, so check with your doctor (or even lawyer) beforehand.

## Building up a tolerance

Now this may sound strange, but some beekeepers (myself included) look forward to getting a few stings early in the season. No, we're not masochistic. The more stings we get, the less the swelling and itching. For many, occasional stings actually build up a kind of tolerance. It still smarts, but the side effects disappear.

One school of thought states that bee venom can actually be good for some health conditions that you may suffer from. This is what bee-sting therapy is all about; see Chapter 1 for more information.

# Understanding Local Laws and Ordinances

Is it legal to keep bees? In most places, the answer is yes. But some areas have laws or ordinances restricting or even prohibiting beekeeping. For the most part, such restrictions are limited to highly populated, urban areas. Other communities may limit the number of hives you can keep, and some require you to register your bees. Ask your local authorities how to go about this. Some communities require that the state bee inspector inspect the health of your colonies periodically. If you have any questions about the legality of keeping bees, contact your state bee inspector, the state department of agriculture, or a local bee club or association.

TIP

*Bee Culture* magazine maintains a terrific online resource under the tab "Resources." This is a great way to find beekeeping clubs, associations, and agencies in your state. Visit www.beeculture.com and follow the links. By the way, there's a subscription discount offer for this journal at the end of this book.

# Easing the Minds of Family and Neighbors

For many among the general public, ignorance of honey bees is complete. Having been stung by wasps and yellow jackets, they assume having any kind of bee nearby spells trouble. Not true. It's up to you to take steps to educate them and alleviate their fears.

Some things you can do to put them at ease are

>> Restrict your bee yard to two hives or fewer. Having a couple of hives is far less intimidating to the uneducated than if you had a whole phalanx of hives.

>> Locate your hive in such a way that it doesn't point at your neighbor's driveway, your house entrance, or some other pedestrian traffic-way. Bees fly up, up, and away as they leave the hive. When they're 15 feet from the hive, they're way above head level.

>> Don't flaunt your hives. Put them in an area where they'll be inconspicuous.

>> Paint or stain your hives to blend into the environment. Painting them flame orange is only tempting fate.

>> Provide a nearby source of water for your bees. That keeps them from collecting water from your neighbor's pool or birdbath (see the "Providing for your thirsty bees" section later in this chapter).

URBAN

Air-conditioning drips are a strong temptation for city bees. Don't let your bees become a nuisance. Be sure to provide your colonies with a nearby source of clean water.

>> Invite folks to stop by and watch you inspect your hive. They'll see firsthand how gentle bees are, and your own enthusiasm will be contagious.

TIP

Keep an extra veil or two on hand for your visiting friends. You certainly don't want a sting on their nose to be what they remember about your bees.

» Let your neighbors know that bees fly several miles from home plate (that's equal to thousands of acres). So mostly they'll be visiting a huge area that isn't anywhere near your neighbor's property.

» Give gifts of honey to all your immediate neighbors (see Figure 3-1 for an example). This gesture goes a long way in the public relations department.

**FIGURE 3-1:**
This gift basket of honey bee products will be given to each of my immediate neighbors. That's sure to help sweeten them up.

**URBAN**

In an urban environment, don't forget your building superintendent and/or the president of your co-op board.

# Location, Location, Location: Where to Keep Your Hives

You can keep bees just about anywhere: in the countryside, in the city, in a corner of the garden, by the back door, in a field, on the terrace, or even on an urban rooftop. You don't need a great deal of space, nor do you need to have flowers on your property. Bees will happily travel a few miles to forage for what they need.

**WARNING**

If you are moving bees to a new location that is a mile or two away, no problem. But if you are moving the hive to a location much less than this, you may lose all of your field bees because they will return to where the hive used to be. If you only need to move your hive a short distance (like across your yard), move the hive a little bit at a time (a yard or two each day) until you reach the desired destination.

# Knowing what makes a perfect bee yard

These girls are amazingly adaptable, but you'll get optimum results and a more rewarding honey harvest if you follow some basic guidelines (see Figure 3-2). Basically, you're looking for easy access (so you can tend to your hives), good drainage (so the bees don't get wet), a nearby water source for the bees, dappled sunlight, and minimal wind. Keep in mind that fulfilling all these criteria may not be possible. Do the best you can by

>> Facing your hive to the southeast. That way your bees get an early morning wake-up call and start foraging early.

>> Positioning your hive so it is easily accessible come honey harvest time. You don't want to be hauling hundreds of pounds of honey up a hill on a hot August day.

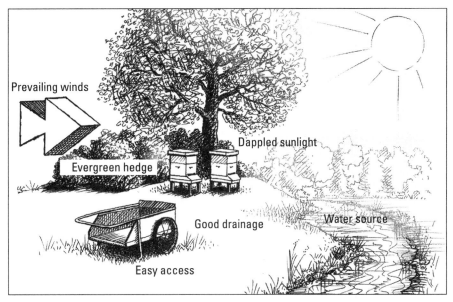

**FIGURE 3-2:** The picture-perfect bee yard. Not always possible, but an admirable objective.

*Courtesy of Howland Blackiston*

>> Providing a windbreak at the rear of the hive (see Figure 3-3). I've planted a few hemlocks behind my hives. Or you can erect a fence made from posts and burlap, blocking harsh winter winds that can stress the colony (assuming you live in a climate with cold winters).

FIGURE 3-3: An ideal backyard location: hemlock windbreak; flat, easy access; dappled sunlight; southeastern exposure; and a nearby water source (not visible in this image).

*Courtesy of Howland Blackiston*

>> Putting the hive in dappled sunlight. Ideally, avoid full sun because the warmth of the sun requires the colony to work hard to regulate the hive's temperature in the summer. By contrast, you also want to avoid deep, dark shade because it can make the hive damp and the colony listless.

TIP

On a city rooftop, fashion a burlap shade roof by stapling the material to four sturdy posts solidly planted in four 5-gallon buckets of cement.

>> Making sure the hive has good ventilation. Avoid placing it in a gully where the air is still and damp. Also, avoid putting it at the peak of a hill, should you live in a region where the bees will be subjected to winter's fury.

>> Placing the hive absolutely level from side to side, and with the front of the hive just slightly lower than the rear (a difference of an inch or less is fine) so rainwater drains out of the hive (and not into it).

TIP

Mulching around the hive prevents grass and weeds from blocking its entrances.

>> Locating your hive on firm, dry land. Don't let it sink into the quagmire.

**URBAN**

If you have doubts regarding the structure of your urban rooftop, check with the building maintenance crew. Don't be famous for being the first urban beekeeper to fall through the roof and into someone's apartment.

# Urban considerations

*(Many thanks to New York City–beekeeper Andrew Cote for his contribution to this section.)*

Just about all the considerations listed in the preceding section apply to urban situations. Here are a few more details for all you city beekeepers out there.

## Dealing with nervous neighbors

Bees living near people can cause anxiety, particularly in the close quarters of city living. In an urban environment, there can be more tension than in the suburbs or in the country. Education is, of course, one of the best remedies for making neighbors more comfortable. And so is a tasty bribe. Give a jar of honey to your neighbors to sweeten their view of bees and beekeeping. Promise them that there's more where that came from!

**TIP**

If your hive has not yet produced honey, don't wait until it does. Just buy someone else's honey and present it to nervous neighbors. Let them know the sweet treat that awaits them in the future.

With the gift jar, include a thoughtfully worded letter that explains what you are planning to do. Let them know it's perfectly legal (assuming it *is* legal in your area). Explain the wonder of honey bees. Exude the virtues of pollination. Share the magic of how local honey fends off pollen allergies. Reassure them that bees rarely sting unless threatened. Give your neighbors time to get past the fear factor; let them digest what they've read and appreciate what they've tasted.

## City bees have the same needs as country bees

Bees in urban settings need morning sun, dappled afternoon shade, a nearby water source (to keep the bees away from air-conditioner drips that are surely nearby), and a buffer from the cold winter wind.

## Deciding where to put your hives

Placement of urban hives is often tricky and a stumbling block for many metropolitan beekeepers. Don't be one of those beekeepers who takes a course, builds a kit, gets a package of bees, and then realizes there's no suitable place to put the bees! Do your homework upfront.

## Strike a deal with a community garden

These are usually run by small neighborhood groups who are sympathetic to honey bees, welcome their pollination, and are likely eager to offer a home for your hives.

REMEMBER

Don't be so grateful for a spot in the community garden that you impulsively offer half of your honey harvest! Your "rent" should be bartered in exchange for the considerable pollination services you bring to the garden. The honey should be yours.

## Speak to your landlord about roof rights

See if you can get access to your building's rooftop and obtain permission to place a hive or two on the roof. Rooftop hives are wonderful in that they are out of sight to most residents, thus reducing neighborhood fear and lessening the chance of vandalism.

Although a roof is a great location for urban beehives, there are safety issues to take into consideration.

>> Avoid a roof if you have to go up a fire escape, climb a tall ladder, or use a rooftop hatch. In all of these situations, it would be difficult (and dangerous) to attempt to remove full and heavy honey supers from the roof area.

>> Don't place your hive too close to the edge of the roof. If you end up with a bee up your pant leg and you lose your balance, no amount of arm flapping will help you fly safely to the ground.

>> Secure all the parts of the hive using *crank straps* (heavy-duty cargo straps with a ratchet locking mechanism). Strong gusts of wind can send hive parts flying wildly off the roof and onto pedestrians below.

WARNING

Never place a beehive on a fire escape. Never. It's illegal, and it's dangerous.

## Providing for your thirsty bees

During their foraging season, bees collect more than just nectar and pollen. They gather a whole lot of water. They use it to dilute honey that's too thick and to cool the hive during hot weather. Field bees bring water back to the hive and deposit it in cells, while other bees fan their wings furiously to evaporate the water and regulate the temperature of the hive.

If your hive is at the edge of a stream or pond, that's perfect. But if it isn't, you should provide a nearby water source for the bees. Keep in mind that they'll seek out the nearest water source. You certainly don't want that to be your neighbor's kiddy pool. You can improvise all kinds of watering devices. Figure 3-4 shows an attractive and natural-looking watering device that I created on top of a boulder that sits in one of my bee yards. All it took was a little cement, a dozen rocks, and a few minutes of amateur masonry skills.

Consider these other watering options: a pie pan filled with gravel and topped off with water, a chicken-watering device (available at farm supply stores; see Figure 3-5), or simply an outdoor faucet that is encouraged to develop a slow drip.

When it comes to providing water for your bees, Figure 3-6 shows a nifty idea that I learned from a fellow beekeeper. Find or purchase a clean pail or bucket. Any size, color, or material will do. Just make sure that it's clean and has never been used for chemicals, fertilizers, or pesticides. Drill ½-inch drainage holes all around the top edge of the bucket. The holes should be placed about 2 to 3 inches down from the top. Fill the bucket nearly to the holes with water, and then float a single layer of Styrofoam packaging pellets on the surface of the water. The pellets give the bees something to stand on as they sip water. That way they won't drown. The drainage holes keep rainwater from overflowing the bucket and washing away the pellets. Neat, huh?

## HOW TO MOVE A FULL HIVE

It's best not to move hives around unless it's necessary because it's disruptive to the bees and a lot of work for you. But sometimes move you must. Here are some helpful guidelines:

- Plan to make your move in the evening when the bees are not flying.

- Before making the move, tape up any extra entrance or ventilation holes you have drilled in the hive (duct tape works great).

- Secure the hive together by using heavy-duty strapping tapes (available at hardware stores). These strapping tapes use a ratchet-type buckle to tighten the straps. Strap the entire hive together as a single unit: bottom board, hive bodies, and cover.

- Tape or gently staple a strip of window screening across the front entrance of the hive. Doing so will keep the bees from flying out of the hive (and stinging you) while providing them with adequate ventilation.

- Use a hand truck to move the hive (an entire hive can weigh a couple hundred pounds). Get some friends to help.

- Wear a veil and gloves in case any bees get loose. I can assure you that they won't be happy about this move.

- Once the hive is in its new location, wait until early the next morning to remove the straps and the entrance screen. This gives the bees time to calm down.

**FIGURE 3-4:**
A shallow bee watering pool that I constructed on a boulder near my hives.

*Courtesy of Howland Blackiston*

**FIGURE 3-5:**
A chicken waterer is a great way to provide your bees with water. Place some gravel or small pebbles in the tray so the bees don't drown.

*Courtesy of Howland Blackiston*

FIGURE 3-6:
The clever
bucket-and-
packing-pellets
solution is a great
do-it-yourself
watering device.

*Courtesy of Howland Blackiston*

**TIP**

You can use a hive-top feeder, filled with water (not syrup) as a convenient way to provide your colony with water.

## Understanding the correlation between geographical area and honey flavors

The type of honey you eat usually is classified by the primary floral sources from which the bees gathered the nectar. A colony hived in the midst of a huge orange grove collects nectar from the orange blossoms — thus the bees make orange-blossom honey. Bees in a field of clover make clover honey and so on. As many different kinds of honey can exist as there are flowers that bloom. The list gets long.

For most hobbyists, the flavor of honey your bees harvest depends on the dominant floral sources in their area. During the course of a season, your bees visit many different floral sources. They bring in many different kinds of nectar. The resulting honey, therefore, can properly be classified as *wildflower honey,* a natural blend of various floral sources.

The beekeeper who is determined to harvest a particular kind of honey (clover, blueberry, apple blossom, sage, tupelo, buckwheat, and so on) needs to locate his or her colony in the midst of acres of this preferred source and must harvest the honey as soon as that desired bloom is over. But doing so is not practical for the backyard beekeeper. Leave it to the professional migratory beekeepers.

**URBAN**

City bees gather a particularly complex plethora of nectars while foraging. That's because most flora planted in urban settings is nonindigenous and immensely varied. This results in a delicious blend of nectars (and thus a delectable honey) that is not found anywhere else on earth.

My advice? Let the bees do their thing and collect from myriad nectar sources. You'll not be disappointed in the resulting harvest, because it will be unique to your neighborhood and better than anything you have ever tasted from the supermarket. Guaranteed!

For a lot more information on honey varietals, see Chapter 15.

# Knowing When to Start Your Adventure

So when do you start your adventure with your bees? The answer depends on where you live. A good time to start is a few months before the "official" launch of the season (when the flowers come into bloom). A chart in Chapter 9 helps you determine the right calendar of events for your region and climate. Generally speaking, in the United States, the season officially starts in the early spring when the bee breeders in the southern states have package bees to sell. Don't wait until the last minute. Use the "winter" months to order and assemble the equipment that you'll need and to reserve a package of bees for early-spring delivery. Read up on bees and beekeeping, and become familiar with your equipment. Join a bee club and attend its meetings. That's a great way to get to know more about beekeeping and meet new friends. Many clubs have special programs for new beekeepers (called "newbees") and hands-on weekend workshops that show you how it's done. Latch on to a mentor whom you can call on to answer questions and help you get started.

Install your bees in the early spring (April or early May is best). Spring varies from area to area, but you're trying to time your start to coincide with the first early-season blossoms, and just a few weeks prior to the fruit bloom. Don't wait until June or July. Starting a hive in summer won't give your colony a chance to grow strong for its first winter.

Be sure to have everything assembled and ready to go before the arrival of your bees. As for what kind of hives and equipment you need to get for this new adventure, that's covered in Chapters 4 and 5.

# Chapter **4**

# Selecting a Hive That's Perfect for You

Beehives come in many different styles, sizes, and shapes . . . many more than I could possibly cover between the covers of this book. So with a little help from my backyard beekeeping colleagues, I've included some hives that would likely be the most popular among the widest range of hobbyist bee-keepers. Each has some benefits and drawbacks, but any one of these is functional and fun for your adventure in beekeeping. This chapter helps you understand the differences among the hives and allows you to select a hive style that best meets your needs and objectives.

I strongly urge the new hobbyist to begin with a Langstroth hive (described in the next section). It's the most widely used hive in the world, and for that reason, the rest of the book focuses a good deal on the use and management of this particular style of hive. However, you'll also find a good deal of detail about the Kenyan Top Bar hive — a hive that is quickly growing in popularity among hobbyist beekeepers. As you gain experience, I also urge you to try some of the other hives mentioned in this chapter. Variety is the spice of life . . . and beekeeping too. You will have a merry time trying your hand at keeping bees in these other styles of hives.

# The Langstroth Hive

Hands down, the Langstroth hive is the most popular and widely used hive today. Certainly this is the case in the United States and in most developed countries around the world. Invented in 1852 by Reverend Lorenzo Langstroth, the basic design has remained mostly unchanged, which is a testament to its practicality. See Figure 4-1.

FIGURE 4-1:
This is a
ten-frame
Langstroth hive,
showing a
bottom board,
two deep-hive
bodies, and
a cover.

*Courtesy of Howland Blackiston*

The big advantage of this hive is that the bees build honeycomb into frames, which can be removed, inspected, and moved about with ease. All the interior parts of this hive are spaced exactly 3/8-inch apart, thus providing the correct *bee space* (in other words, the bees won't glue parts together with propolis or fill the space with burr comb).

Given its wide popularity, many commercially available parts and accessories sold for beehives are standardized to accommodate this design, meaning the beekeeper using a Langstroth hive has all kinds of options when it comes to purchasing extras (such as replacement parts and accessories).

Consider standardizing on using a *medium* honey super (the super is where the bees store the surplus honey you will harvest). Many beekeepers use *shallow* supers, which are not quite as tall as the medium design. But I have a strong preference for using medium honey supers rather than shallow. Why? Because they hold more honey than shallow supers, and yet they are still relatively small and light enough to lift off the hive.

Many suppliers are offering an eight-frame version of the Langstroth hive. Fewer frames result in a setup that's lighter in weight, and that can be a big advantage to the beekeeper when lifting hive bodies. Your back will appreciate the eight-frame version! This is an option that's gaining great popularity among backyard beekeepers.

TIP

The Langstroth is the hive type that I recommend to new beekeepers (see Figure 4-2).

FIGURE 4-2:
This is an eight-frame version of the traditional Langstroth hive, sometimes marketed as the Garden Hive.

*Courtesy of Brushy Mountain Bee Farm*

Some of the advantages to consider regarding the Langstroth hive include

>> **Capacity:** Because this design consists of modular, interchangeable hive parts, you can add extra medium or shallow honey supers as the colony grows and honey production increases. Capacity for bees and honey is virtually unlimited.

>> **Frames:** There are both eight- and ten-frame versions of the Langstroth hive available (the eight-frame version being a slightly smaller and lighter hive in weight than its ten-frame cousin). Both versions use a Langstroth-style of self-centering frame with beeswax foundation inserts. Chapter 5 includes information on assembling these frames and foundation.

>> **Universality:** Because the Langstroth hive is so widely used around the world, you can easily find replacement parts, gadgets, and add-ons for this popular hive. They are widely available from many beekeeping supply stores (search the web and you will find many such suppliers).

TIP

Appendix A includes a list of some suppliers; check out the back of this book for some valuable discount coupons and special offers from these vendors.

# The Kenyan Top Bar Hive

In recent years, there has been increasing interest in this elegantly simple, practical design among backyard beekeepers, particularly because more and more backyard beekeepers are seeking natural and environmentally sustainable options. Top Bar hives are used extensively throughout Africa and in other developing countries where building materials are scarce. This particular design (shown in Figure 4-3) is referred to as a Kenyan Top Bar hive because of its sloped sides that tend to result in stronger, more balanced combs. In addition, the sloped sides discourage the bees from attaching the comb to the bottom and sides of the hive.

**FIGURE 4-3:** The Kenyan Top Bar hive is likely one of the oldest beehive designs, going back centuries.

*Courtesy of Bee Thinking*

There are many design variations to a Kenyan Top Bar hive. Basically they all consist of a long, horizontal hive body with sloped sides, top bars (not frames), and a roof. It's a very practical and functional hive. But like all Top Bar hives, there are some pros and cons. First, the pros:

- » The top bars (which have no bottom rails, no side rails, and do not use full sheets of foundation) allow the bees to build their comb naturally and without foundation as a guide for cell size.

- » The design provides a habitat that is similar to how bees live in the wild (within the hollow of a tree). As a result the bees exhibit more natural behaviors.

>> Although a growing number of vendors sell this style of hive, the hive is relatively easy to build yourself at low cost.

>> No honey extractor is needed because the comb is typically cut from the hive and crushed to extract the honey. Alternatively, you can harvest chunks of natural comb honey.

>> There is less heavy lifting than a conventional-style hive, where supers and hive bodies need to be removed and/or manipulated (see Figure 4-4).

>> There is no need to store empty honey supers during the winter months because there are none.

>> The hive provides a great source of chemical-free beeswax because the comb containing honey is removed during the honey harvest. The wax can then be used by the beekeeper for making candles, cosmetics, and furniture polish (for recipes and information on the products you can make from beeswax, see Chapter 18).

**FIGURE 4-4:**
Here's a top bar with beautiful, all-natural comb, heavy with honey, pollen, and brood. It doesn't get much better than this!

*Courtesy of William Hesbach*

**ALL NATURAL**

The Kenyan Top Bar hive is easy and inexpensive to make from recycled scrap wood. It's about as close as you can get to providing your bees with a natural living environment.

However, there are also some cons:

>> Harvesting of honey requires destroying the comb (it must be crushed to extract the honey from the comb). On the other hand, it means that every year, the colony uses a fresh supply of beeswax to build new honeycomb, and there is less chance of pesticide buildup that can be found in wax that is used over and over, year after year.

>> The honey harvested may contain a higher percentage of pollen and be cloudy in appearance. But one could argue this is far more natural and organic and provides the honey with extra protein (pollen content). That cloudiness is a healthy problem to have!

>> Making artificial swarms, splits, reversals, and other manipulations, such as feeding sugar syrup, are difficult or even impossible (versus what's possible with a hive using framed comb); however, the whole idea of this design is to provide as natural an environment as possible for your bees. Frequent manipulations are contrary to the all-natural objective.

>> It can be more difficult to administer medications. But with a Top Bar hive, you're practicing a more natural approach to beekeeping, allowing for less or no medication and chemicals. Bees tend to be healthier when under less stress and in a more natural environment. And this Top Bar hive provides just that.

>> There tends to be lower honey production from this type of hive.

>> Combs are much more fragile than comb built in a Langstroth-style frame (with the support of side bars and a bottom rail). Care needs to be taken when removing and handling combs of a Top Bar hive.

>> Some feel this horizontal design may not be the best option for those keeping bees in areas with long and very harsh winters; however, I know a number of beekeepers successfully using this design here in New England (where we have plenty of winter).

>> This hive has the largest footprint of the hives mentioned in this chapter. That may be an issue for beekeepers with limited real estate to work with (such as urban beekeepers).

**TIP**

A degree of special knowledge is needed to manage this hive, and in this edition I provide new information that will be of value to those interested in beekeeping with a Kenyan Top Bar hive. Also the Internet has a growing number of useful articles that are helpful to beekeepers interested in trying this time–tested design. And you will even find helpful user forums on social media sites.

The qualities and characteristics of the Kenyan Top Bar hive will attract some beekeepers and turn others away. Here are some of the basics to consider:

» **Capacity:** There is no possibility to add supers or additional hive bodies. The space is fixed and thus limited to the size of the hive and the top bars it accommodates.

» **Frames:** There are no frames. This design uses top bars. There are no side bars, no bottom bar, and no full sheets of beeswax foundation to deal with (nor the costs associated with these elements). The bees attach their comb naturally onto each of the top bars placed in the hive. The Top Bar hive can be made any length, but it's typically between 3 and 4 feet long, containing approximately 28 top bars in all.

» **Universality:** This statistic becomes less relevant with this hive type. The dimensions for Top Bar hives are not standardized, and thus the gadgets and add-ons that are easy to find for Langstroth hives (for example, feeders, screened bottom boards, honey-extracting equipment) are usually not commercially available for Top Bar hives. Often the beekeeper must custom build his own accessories.

**REMEMBER**

If you plan to also have Langstroth hives, none of the equipment is interchangeable between Top Bar and Langstroth hives.

# The Warré (People's) Hive

The Warré hive (shown in Figure 4-5) is also a Top Bar–style hive, which was developed in France during the early 20th century by Abbé Émile Warré, an ordained priest and an avid beekeeper. His vision was to develop an easy-to-build and easy-to-manage beehive with which everyone could have success (thus, it is often referred to as *the People's hive*). This simple, cost-effective, and efficient design has been gaining renewed popularity among DIY beekeepers and those seeking more *organic* approaches to beekeeping. I appreciate the more natural conditions that the hive provides the bees.

» The top-bar frames (which have no bottom rails, no side rails, and don't use full sheets of foundation) allow the bees to build their comb freestyle and without restrictions on cell size.

» The design provides a living arrangement that is similar to how bees live in the wild (within the hollow of a tree). As a result, the bees tend to be less stressed and thus less prone to disease.

>> The vertical design allows the bees to grow the colony as they do in the wild (working downward from top to bottom).

>> Beekeepers using this design tend not to disturb the colony with frequent swarm-prevention inspections, so there is excellent retention of the colony's scent. In addition, a hive that is not opened often by the beekeeper gives the colony much better natural control of temperature and humidity.

>> With plenty of space to grow the brood nest, the risk of swarming is greatly reduced.

>> The hive yields lots of beeswax because most of the comb is removed during the honey harvest. The wax can then be used by the beekeeper for making candles, cosmetics, and furniture polish. (For recipes and information on the products you can make from beeswax, see Chapter 18.)

**FIGURE 4-5:**
The Warré hive is a Top Bar hive without foundation that allows the bees to build comb as they would in nature.

*Courtesy of Bee Thinking*

**ALL NATURAL**

The Warré Top Bar hive is a favorite among beekeepers practicing a *natural* approach to beekeeping (see Figure 4-6).

**FIGURE 4-6:** These are the top bars used in a Warré hive. Note there are no side or bottom bars, and no wax foundation. The bees build their own comb freely onto these bars.

**WARNING**

However, with the good comes some bad. Here are some of the negatives associated with the design:

>> The capacity of a Warré hive is small, thus swarming may become commonplace.

>> Because the Warré colony grows by your adding hive boxes *below* (not *above* as with a Langstroth), a good deal of heavy lifting is required.

>> Harvesting of honey requires destroying the comb (it must be crushed to extract the honey from the comb). On the other hand, it means that every year the colony uses a fresh supply of beeswax comb, and there is less chance of pesticide buildup that can be found in wax that is used over and over, year after year.

>> The honey harvested may contain a higher percentage of pollen and be cloudy in appearance. But one could argue this is far more natural and organic, providing the honey with extra protein (pollen content), and thus can be considered healthier.

>> Making artificial swarms, splits, reversals, and other manipulations, such as feeding sugar syrup, are difficult or even impossible (versus what's possible with a hive using framed comb). But the whole idea of this design is to provide as natural an environment as possible for your bees. Frequent manipulations are contrary to the all-natural objective.

>> If you decide to administer medications, that can be a difficult task (there is no feeder, for example). But hey, with a Warré hive, we're talking all-natural, sustainability, and organic beekeeping. Medication and chemicals need not apply at all. The great news is that bees tend to be healthier when under less stress and in a more natural environment. And this simple design provides just that.

TIP

Keeping bees in a Warré hive requires a different skill set than raising bees in a Langstroth hive. Search the Internet for articles and books about keeping bees in a Warré hive.

Here are some of the reasons that the Warré hive has gained such popularity among hobbyists:

>> **Capacity:** The tall, vertical design provides ample room for the colony to naturally grow throughout the season. There is no need for the beekeeper to anticipate the colony's growth and add hive bodies during the season. All the space that's needed is provided from the get-go.

>> **Frames:** This design does not use frames or foundation. It uses top bars, upon which the bees draw out their own natural comb. There are no side bars, no bottom bar, and no full sheets of beeswax foundation to deal with (nor the costs associated with these elements). The bees build their comb without restriction onto each of the top bars placed in the hive (typically there are 32 top bars in all).

>> **Universality:** This statistic becomes less important with this design. All the gadgets and add-ons that you might use with a conventional Langstroth hive (feeders, queen excluders, foundation, and honey-extracting equipment) are irrelevant to the Warré. Its design is virtually all-inclusive and requires no extras; however, if you need a new roof, there are an increasing number of beekeeping supply stores offering pre-built Warré components. Just make sure the dimensions of the commercially made hives jive with your hive's dimensions (there is less standardization regarding Warré measurements than with the more-popular Langstroth hive).

# The Flow Hive

The Flow hive, shown in Figure 4-7, is an award-winning invention of Australian father-and-son team Stuart and Cedar Anderson. They invented what is essentially a patented way of harvesting the surplus honey from the hive without having to smoke the bees and remove the honey supers. There is no slicing cappings off the comb, no spinning the frames, nor is there a need for an extractor. A significant amount of labor has been eliminated from the process of harvesting honey.

FIGURE 4-7:
The Flow hive introduces a new way to harvest honey, by simply turning a lever on the hive.

*Courtesy of BeeInventive*

The secret is in the Flow Frames, which have been specially engineered to surrender the capped honey with just a turn of a lever called the Flow Key. The key opens a channel within the honeycomb, and the honey drains to a pipe and directly into your container of choice. Meanwhile, back in the hive, the bees are virtually undisturbed as the honey drains from under their feet. When you finish draining, you just turn the lever back and the cells are reset and ready to be refilled.

The technology is clearly a breakthrough from the way we have been traditionally harvesting honey for centuries. This has created enormous interest across social media and stimulated the sale of thousands of these hives. It's only been on the market since 2015, so as of this writing, there has only been a season of widespread use. But I for one am optimistic about this product, and enthusiastic that creative people have come up with a new breakthrough in beekeeping.

REMEMBER

What seems critical to me (a point which the company stresses) is that new beekeepers understand that the Flow hive is not a shortcut to responsible beekeeping. Yes, it provides a fun and really easy way to harvest honey, but it does not mean you can just pop this in your yard and simply wait to harvest honey. You must install a package of bees and manage and care for your bees just as you would with any other hive. You must routinely inspect the colony and take action when your help is needed. There are no shortcuts to being a good beekeeper, but with the

Flow hive, you may just have a bit of fun and save some time when it comes to harvesting your honey!

Here are the pros and cons of using the Flow hive, starting with some pros:

>> The technology does greatly simplify the honey harvesting process and reduces much of the labor associated with conventional honey harvesting.

>> The design means you don't have to smoke and open your hives, remove heavy honey supers, and otherwise expose yourself to the possibility of retaliatory stings.

>> The Flow Frames are designed to fit Langstroth-style deep boxes (although they will adapt to fit other hive types). The Flow Frames are inserted in much the same way as standard conventional frames. The deep box is modified slightly to facilitate the Flow Frames operation. You can modify your own Langstroth deep-hive box or purchase an already-modified box from the company.

>> To its credit, the company's website (www.honeyflow.com) has a tremendous amount of information and helpful videos about the Flow hive and its setup, operation, and use in a wide variety of hive types.

>> The honey extracted from the Flow Frames is far "cleaner" than honey extracted the traditional way, meaning there is less debris (wax bits) to be filtered from the honey.

>> Different nectar is available at different times, so each Flow Frame tends to fill up from a single nectar source. This results in individual frames storing different honeys (from different nectars). Because the frames are clear plastic, you can see the different colors of honey from frame to frame. Because you can harvest and bottle the honey one frame at a time, the Flow hive allows for the tasting of different honey varietals. Yummy.

And with this design come a few cons:

>> The Flow Frames are far more expensive that conventional frames, although you must consider you really don't need to invest in an expensive extractor or other traditional harvesting equipment. By the way, other than your honey super with these frames, the other deeps (for brood) use conventional deep frames and foundation (or conventional plastic frames and foundation if that's your preference).

>> You will not be able to harvest the wax capping from these honey frames.

>> Granulated honey cannot be extracted from the Flow Frames. But then again, granulated honey can't be extracted using traditional harvesting methods.

>> If you don't follow the precautions stipulated on the Flow website, you can inadvertently set off a robbing frenzy or attract opportunistic pests, such as ants, skunks, raccoons, and so on.

>> New beekeepers must understand that this new invention is not an excuse to bypass the responsibilities of a good beekeeper. You must still perform all the other tasks of stewardship that are outlined in this book. This point is not a criticism of this hive, but more of a warning that as easy as this hive makes honey harvesting, it does not mean it is any easier when it comes to your other tasks.

Here are the basic stats for the Flow hive:

>> **Capacity:** You can add extra Flow Frames and supers as the colony grows and honey production increases. Capacity is virtually unlimited.

>> **Frames:** The Flow Frames are what make the technology work. These are specially engineered frames only available from the manufacturer. The "magic" won't happen unless you use these frames. As mentioned before, the frames readily fit in a modified Langstroth deep box, and other hive types too (see the Flow website for more detail regarding compatibility with other hive types).

>> **Universality:** Although you can place both standard deep frames and foundation and Flow Frames in the same hive box, only the Flow Frames will allow you to harvest honey with a turn of the key. And at present, only this company manufactures the Flow Frames.

# The Five-Frame Nuc Hive

The design of the five-frame nuc hive is similar to the Langstroth hive, only in miniature (see Figure 4-8). Although the nuc hive may not be your only hive in the yard, it's mighty handy to have one or two of these nearby. Why this mini-me hive? Some of the reasons include the following:

>> A nuc can serve as a nursery for raising new queens (to learn how to raise queens, see Chapter 14).

>> A nuc provides you with a handy source of brood, pollen, and nectar to supplement weaker colonies (kind of like having your own dispensary).

>> A nuc colony of bees can be sold to other beekeepers — they're a fast way to populate a new hive.

>> A nuc can be used to populate an observation hive.

» A nuc can be used to house a captured swarm of bees.

» A nuc can be used to house a supply of bees used for bee-venom sting therapy. You can find more information on the use of bee-venom to treat certain inflammatory medical conditions from the American Apicultural Society (www.apitherapy.org).

» A nuc in the corner of a garden can help with pollination, and it can require far less maintenance than a regular hive. (Note, however, there will be no harvesting of honey from this tiny little hive.)

FIGURE 4-8:
A nuc hive (sometimes called a nucleus hive) is a small hive with a few frames of bees.

*Courtesy of Bee-Commerce*

A disadvantage of a nuc hive is that it doesn't overwinter well in colder climate zones. It *can* be done, but it takes effort and talent on the part of the beekeeper. There are just not enough bees or stored honey available to see the little colony through the winter months. So if you live in an area where the winters are cold (long periods below freezing), you should combine your nuc colony with one of your big hives before Jack Frost pays a visit. However, if you have mild winters, you can feed the nuc colony using an entrance feeder, and they should do fine.

Another disadvantage is that these small hives are simply more fragile and difficult to manage than larger hives. They get overcrowded and swarm frequently, and the population is too small to deal with environmental ups and downs.

In short, nuc hives do require a bit more skill on the part of the beekeeper to care for.

They say that good things come in small packages. See what you think as you review the basic statistics for the five-frame nuc:

» **Capacity:** Because this design consists of only five frames, there is no room for expansion as the colony grows in population, so the capacity for bees is limited. Theoretically, you could build additional nuc hive bodies and stack them one on top of the other to allow the colony to grow. That's not really the purpose of a nuc hive, however, and I've rarely seen this done.

» **Frames:** This nuc hive uses a Langstroth-style of self-centering deep frame with deep beeswax foundation inserts. A total of five frames are used with the nuc hive.

» **Universality:** Because the nuc uses Langstroth-style frames, you can easily purchase the right size foundation for this hive. Virtually all suppliers sell nuc hives and their associated parts.

# The Observation Hive

An observation hive (shown in Figure 4-9) is a small hive with glass panels that enable you to observe a colony of bees without disturbing them or risking being stung. Such hives are usually kept indoors but always provide access for the bees to fly at will through a tube from the hive to the outdoors. The entry can be opened or closed at the beekeeper's whim. These hives also have a means for feeding the bees (sugar syrup) to supplement what they are able to store in this confined space.

**FIGURE 4-9:** Here are two different styles of observation hives: The thicker one on the left is a six-frame hive (containing three rows of two frames). The one on the right is a thinner two-frame hive (with room for only a single row of two frames). The more frames, the more likely you can overwinter the colony.

*Courtesy of Bee-Commerce and Dadant*

With an observation hive, you must allow the bees their freedom to fly outside to relieve themselves as well as to gather pollen, nectar, and water. So although you can close the entry and take them to an event where they can be displayed, the hive must not be kept closed up (that is, the bees not able to fly) for more than 6 to 12 hours.

I'm a big believer in having an observation hive — even when you have conventional hives in your garden. The pleasure and added insight they give you about honey-bee behavior is immeasurable. You learn a lot about bees when you can see what's going on in the colony.

Among a few of the rewards you can realize from setting up an observation hive are that it

» Serves as a fantastic educational tool to take to schools, farmers' markets, garden shows, and so on. The smaller ones (fewer frames) are very light-weight and portable.

» Gives you a barometer on what's happening in a bee colony at any given time of year. That way you can anticipate the needs of your outdoor colonies and better manage your hives.

» Makes possible safe, close-up observation of bee behavior. And because you can watch the bees without smoking or opening the hive, the bees' actions are far more natural. You see things that you can never witness while smoking and inspecting a conventional hive.

» Provides (because it's kept *indoors*) year-round enjoyment. No need to be a seasonal beekeeper, because you can observe your bees even in the dead of winter.

» Enables you to enjoy the pleasures of beekeeping from the comfort of your home, especially when you don't have the space to keep bees outdoors or can't physically manage a robust outdoor hive.

Observation hives are a great choice for urban beekeeping, providing you are not keeping bees for pollination services or honey harvesting.

But with all the benefits come a few disadvantages:

» Because of the fixed size of this hive, there's no room for the colony to grow. You can expect a colony kept in an observation hive to swarm at least once a season, regardless of the hive's size.

» There is no honey harvested from an observation hive.

>> The population of an observation hive is not significant enough to provide any meaningful pollination to your garden.

>> You need to continually feed sugar syrup to the bees. There is not enough room in the hive for the bees to store significant amounts of honey and pollen. Most observation hives have a mechanism for feeding the colony syrup without opening the hive.

Observation hives come in all sizes and styles. Some contain a mere frame or two. These smaller observation hives are great for toting to garden shows, classrooms, or wherever you might do a "show and tell" about beekeeping. But these small observation hives don't have enough volume for housing a decent-sized colony that can survive throughout the season or during the winter months.

**TIP**

Keeping bees in an observation hive requires a different skill set than raising bees in most other hives. For detailed information about setting up, maintaining, and using an observation hive, I recommend the book *Observation Hives* by Thomas Webster and Dewey Caron (A.I. Root Company).

Here are some things to consider as you evaluate whether an observation hive is your cup of tea:

>> **Capacity:** This varies depending on the style of observation hive you purchase. Figure 4-9 shows two different styles — two-frame and six-frame capacity. Regardless of how many frames they have, observation hives do not typically allow room for expansion as the colony grows in population, and thus the capacity for bees is limited. So prepare yourself for the inevitable — they're going to swarm! Now that makes for a dramatic event!

>> **Frames:** Most commercially available observation hives come with a Langstroth-style of self-centering frame, although there's no reason you couldn't try using top bars in your observation hive.

>> **Universality:** There is no single standard design for an observation hive, so universality does not apply.

# Make a Beeline to the Best Beehive

So with all the different hives to choose from, how do you decide? Maybe you just like the look of one hive over another. A better way to decide is to determine the primary reason you are beekeeping and select the hive best suited for that reason. Table 4-1 at the end of this section will help you make these decisions.

If you are a new backyard beekeeper just learning about bees and how to manage them, the Langstroth hive (either eight- or ten-frame version) is your best bet. After a year or two of experience, try your hand at some of the other hives mentioned in this chapter. They require different techniques when it comes to managing your colonies. Many good resources are available on the Internet for managing Warré, Kenyan Top Bar, nucs, and observation hives.

The following sections take a look at the various things that might be going through your mind as you think about which of these hives you would like to try.

## Hives for harvesting honey

You will be able to harvest from most of these hives, except for the nuc and observation hives. Both the Warré and the Kenyan Top Bar hives are decent honey producers. And the Flow hive makes harvesting honey a breeze. But if you want gobs and gobs of honey, the Langstroth is the hive for you. It will have the largest colony of bees busily making honey. And its modular design allows you to add as many honey supers as the season dictates. Rest assured, the Langstroth is the granddaddy of all honey producers.

## Hives for pollinating your garden

Suppose your primary reason for having bees is to improve pollination in your garden. You don't care about harvesting honey. You don't care about a show-and-tell hive. You just want larger and more abundant flowers, vegetables, and fruits.

The good news is that any of these hives will help pollinate a garden. But they will accomplish this to varying degrees. The larger the hive, the larger the colony to carry out pollination. Also, the larger the hive, the more work for you. So, if your intent is to max out pollination, then consider the Kenyan, Warré, Flow, or Langstroth hives. If you don't want all the work associated with larger hives, a nuc hive tucked into the corner of your garden or fruit orchard will likely do a reasonably respectable job of pollinating.

## A hive for learning and teaching

Suppose you are really interested in bees but don't want to deal with all that outdoors stuff every week. Your primary interest is to learn more about bees. To study their behavior. To sit back and observe the fascinating things that take place inside a hive. Kind of like having an aquarium where one becomes mesmerized by watching the fish do their thing. You feed them a little and occasionally clean the glass, but that's about it. Or, you want to make presentations at schools, nature centers, and farmers' markets, and you need a hive that's portable and can be used to safely display live bees.

Then the best choice for you is the observation hive. Choose any of the sizes or styles that suit your fancy. It's a way to show and tell and to, well, observe.

Table 4-1 can help you decide which of the various hives mentioned in this book are just right for your adventure in beekeeping.

**TABLE 4-1** **Hive Checklist — Which Hive Meets Your Needs?**

| Hive Type | Show and Tell | Pollination | Honey Extraction | Notes |
|---|---|---|---|---|
| Langstroth | No | Good | Very Good | This is my recommended hive for someone getting started with beekeeping. Select either the eight- or ten-frame versions. |
| Kenyan Top Bar hive | No | Good | Good | A great hive for those seeking a natural environment for their bees. The overall maximum size of the hive is fixed, so there is no opportunity to add supers and thus increase honey yields beyond what's possible with the given size of the hive. |
| Warré hive | No | Good | Good | Another great hive for those seeking a natural environment for their bees. It has the smallest footprint of all the exterior hives in this book, which makes it ideal for someone with limited space. |
| Flow hive | Yes | Good | Very Good | Some of these hives come with "windows" that allow you to observe what's going on inside, but the hives are not "portable" for traveling to events. The easy harvesting of honey is the big selling feature of this unique hive. |
| Nuc hive | No | Moderate | N/A | This is a very small hive. It's ideal for starting a new colony, raising queens, or providing some pollination in a garden. |
| Observation hive | Yes | Poor | N/A | This is the only hive in this group that is portable and lightweight enough to travel to various teaching opportunities. It's also the only hive that allows you to observe bee behavior 24/7. |

# THE SUN HIVE

Here's a design so interesting and inspiring, I just couldn't resist sharing it.

The sun hive (Weissenseifener Haengekorb) was designed by the German sculptor Guenther Mancke. This design started appearing in the early 1990s. It is based on the simple basket hive, or *skep,* that was popular for hundreds of years in many countries. (The skep is associated with the public's "romantic" image of what a beehive looks like.)

The sun hive's form and shape was inspired by Mancke's study of natural/feral bee nests. His design includes an ingenious combination of skep baskets (woven from rye straw) that are often covered with an insulating plaster made from cow dung. Inside there are nine half-moon-shaped, movable, wooden frames. The bees build their comb naturally onto the frames (like a Top Bar hive). The entrance is located at the bottom of the hive.

Unlike the picturesque straw skeps of yesteryear (which are illegal to use in the United States), the sun hive has removable frames that can be inspected. It's this design feature that theoretically should make the sun hive okay to use, but its legality in your area should be validated by the powers that be.

If you want one of these beauties, you will likely have to make one yourself (check the Internet for how-to-build workshops in the United States and Europe). Few people are making these hives for sale, and even if you do find a builder, the waiting list would likely be long.

Courtesy of Michael Joshin Thiele: www.gaiabees.com.
Photographer Amanda Lane: www.amandalane.com

# Chapter **5**

# Basic Equipment for Beekeepers

Beekeepers use all kinds of fantastic tools, gadgets, and equipment. Quite frankly, part of the fun of beekeeping is putting your hive together and using the paraphernalia that goes with it. The makings for a beehive come in a kit form and are precut to make assembly a breeze. The work is neither difficult nor does it require too much skill. Some suppliers will even assemble the kits for you.

**REMEMBER**

The more adventuresome among you may want to try making your own hives from scratch. After you get the hang of beekeeping, you can try your hand at making your own hives. Many different mail-order establishments offer beekeeping supplies, and several excellent ones are now on the Internet. Check out a listing of some of the quality suppliers in Appendix A.

**URBAN**

If you're planning to place a hive on the roof of an urban building, make sure that all the parts of the hive style you choose will fit through the window/trap door/hatch that allows you access to your roof.

# Starting Out with the Langstroth Hive

Many different kinds and sizes of beehives are available. Check Chapter 4 for the stats on several of the most popular hives for the backyard beekeeper. But worldwide, the most common is the eight- or ten-frame *Langstroth hive.* This so-called moveable-frame hive was the 1851 invention of Rev. Lorenzo L. Langstroth of Pennsylvania. Its design hasn't changed much in the last 150 years, which is a testament to its practicality. Therefore, this is the style of hive that most of this book is devoted to. However, because of the increasing popularity of the Kenyan Top Bar hive, this edition includes supplemental information regarding it as well.

Here are some of the benefits of the Langstroth hive:

>> Langstroth hive parts are completely interchangeable and readily available from any beekeeping supply vendor.

>> All interior parts of the hive are spaced exactly ⅜ inch (9.525 mm) apart, thus enabling honey bees to build straight and even combs. Because it provides the right "bee space," the bees don't "glue" parts together with propolis or attempt to fill spaces with burr comb.

>> Langstroth's design enables beekeepers to freely inspect and manipulate frames of comb. Before this discovery, beekeepers were unable to inspect hives for disease, and the only way to harvest wax and honey was to kill the bees or drive them from the hive.

>> The traditional Langstroth hive body holds ten frames, but an eight-frame version is readily available. The advantage is that eight-frame hives are a bit less cumbersome and less heavy than the ten-frame original.

**TIP**

Decide upfront whether you will use the eight- or ten-frame variety of the Langstroth hive. Standardize on one or the other, because the hive bodies are not interchangeable with each other. Pick the ten-frame version for maximum honey production. Pick the eight-frame version for lighter and easier-to-carry hive bodies and honey supers. More and more hobbyists are settling on the lighter eight-frame option. Those ten-frame supers get super heavy!

# Knowing the Basic Woodenware Parts of the Langstroth Hive

*Woodenware* refers to the various components that collectively result in the beehive. Traditionally these components are made of wood — thus the term — but some manufacturers offer synthetic versions of these same components (plastic, polystyrene, and so on). My personal preference: Real wood. The bees accept it far more readily than synthetic versions. And the smell and feel of wood is ever so much more pleasurable to work with.

TIP

Be aware that the hive parts you order (see "Ordering Hive Parts" later in this chapter) will arrive in precut pieces. You will need to spend some time assembling them. See "Setting up shop" later in this chapter for a list of tools and hardware that you will need for assembly. *Note:* Some vendors will preassemble hives for you, usually for a slight extra charge.

This section discusses, bottom to top, the various components of a modern ten-frame Langstroth beehive. As you read this section, refer to Figure 5-1 to see what the various parts look like and where they are located within the structure of the hive.

## Hive stand

The entire hive sits on a hive stand. The best ones are made of cypress or cedar — woods that are highly resistant to rot. The stand is an important component of the hive because it elevates the hive off the ground, improving circulation and minimizing dampness. In addition, grass growing in front of the hive's entrance can slow the bees' ability to get in and out. The stand helps alleviate that problem by raising the hive above the grass.

TIP

The hive stand consists of three rails and a landing board, upon which the bees land when they return home from foraging trips. Nailing on the landing board just right is the only tricky part of hive stand assembly. Carefully follow the instructions that come with your hive stand. *Note:* Putting the stand together on a flat surface helps prevent the stand from wobbling.

URBAN

When placing a beehive on a roof, the surface will often be unsteady. It's a good idea to bring a few shims (slats of wood) to steady the hive. Also, don't neglect a hive stand just because there is no grass on the roof. Here's another hint: Save your back the trouble and place the hive on an *elevated* hive stand that raises the entire hive setup 12 inches or more off of the roof's surface. There's more about elevated hive stands later in this chapter.

## Bottom board

The *bottom board* is the thick bottom floor of the beehive. Like the hive stand, the best bottom boards are made of cypress or cedar wood. This part's easy and intuitive to put together.

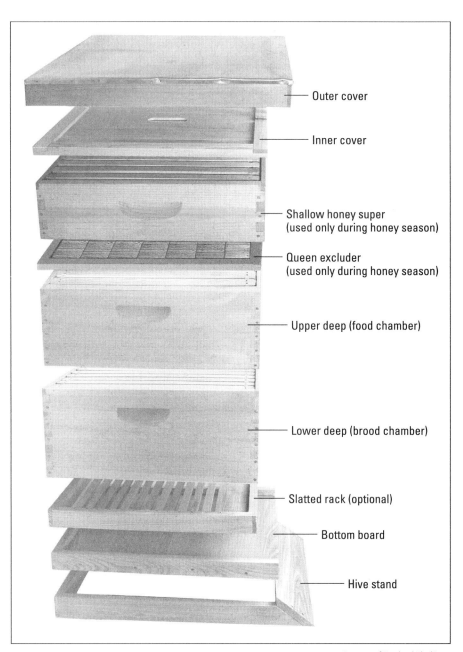

**FIGURE 5-1:**
The basic components of a modern Langstroth hive.

- Outer cover
- Inner cover
- Shallow honey super (used only during honey season)
- Queen excluder (used only during honey season)
- Upper deep (food chamber)
- Lower deep (brood chamber)
- Slatted rack (optional)
- Bottom board
- Hive stand

*Courtesy of Howland Blackiston*

**TIP**

Some beekeepers use what's called a *screened* bottom board in place of the standard bottom board. This improves ventilation and is helpful when monitoring the colony's population of varroa mites (see Chapter 13 for more information on varroa mites).

# Entrance reducer

When you order a bottom board, it comes with a notched wooden cleat. The cleat serves as your *entrance reducer*, which limits bee access to the hive and controls ventilation and temperature during cooler months. The entrance reducer isn't nailed into place, but rather is placed loosely at the hive's entrance. The small notch reduces the entrance of the hive to the width of a finger. The larger notch (if available on the model you purchase) opens the entrance to about four finger widths. Removing the entrance reducer completely opens the entrance fully.

Beekeepers use the entrance reducer only for newly established hives or during cold weather (see Chapter 6). This is the reason the entrance reducer isn't shown in Figure 5-1. For established hives in warm weather, the entrance reducer isn't used at all. The only exception may be when you're dealing with a robbing situation — see Chapter 10.

TIP

If you can't find your entrance reducer, use a handful of grass to reduce the hive opening.

# Deep-hive body

The *deep-hive body* typically contains eight or ten frames of honeycomb (depending on the type/size of Langstroth hive you are using). The best quality frames have crisply cut dovetail joints for added strength. You'll need two deep-hive bodies to stack one on top of the other, like a two-story condo. The bees use the *lower deep* as the nursery, or *brood chamber,* to raise thousands of baby bees. The bees use the *upper deep* as the pantry or *food chamber,* where they store most of the honey and pollen for their use.

TIP

If you live in an area where cold winters just don't happen, you don't need more than one deep-hive body for your colony.

The hive body assembles easily. It consists of four precut planks of wood that come together to form a simple box. Simply match up the four planks and hammer a single nail in the center of each of the four joints to keep the box square. Use a carpenter's square to even things up before hammering in the remaining nails.

Place the hive body on the bottom board. If it rocks or wobbles a little, use some coarse sandpaper or a plane to remove any high spots. The hive body needs to fit solidly on the stand.

TIP

Use a little waterproof wood glue on the joints of all your woodenware before nailing them together. That gives you a super-strong bond.

If you're handy, you may want to try building your own equipment. For the adventuresome, I include plans in Chapter 18 for building a hive stand. For the supremely adventuresome, check out my book, *Building Beehives For Dummies* (Wiley). It has detailed how-to instructions for making a range of popular beehives and beekeeping accessories from scratch.

Remember that precise measurements are critical within any hive. Bees require a precise bee space. If you wind up with too little space for the bees, they'll glue everything together with propolis. Too much space, and they'll fill it with burr comb. Either way, it makes the manipulation and inspection of frames impossible. So, as the old saying goes, measure twice and cut once!

## Queen excluder

No matter what style of honey harvest you choose, a queen excluder is a basic piece of equipment that many backyard beekeepers use. It's placed between the deep food chamber and the shallow (or medium) honey supers, the parts of the hive that are used to collect surplus honey. The queen excluder comes already assembled and consists of a wooden frame holding a grid of metal wire or a perforated sheet of plastic (see Figure 5-2). As the name implies, this gizmo prevents or excludes the queen from entering the honey super and laying eggs. Otherwise, a queen laying eggs in the super encourages bees to bring pollen into the super, potentially spoiling the clarity of the honey. The spacing of the grid is such that smaller worker bees can pass through to the honey supers and fill them *only* with honey.

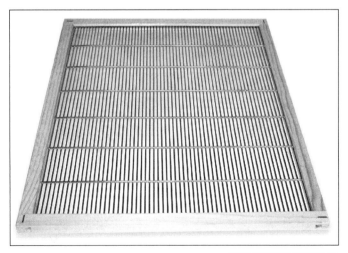

**FIGURE 5-2:**
A queen excluder.

*Courtesy of Howland Blackiston*

**WARNING**

You use a queen excluder only when you place honey supers on the hive and the bees are bringing in nectar and making it into honey. It is a piece of woodenware that is unique to honey production. When you are not collecting honey, it should not be used.

Many experienced beekeepers (myself included) will not use a queen excluder because doing so can slow down the bees' progress in producing honey. Some say it may even contribute to swarming (see Chapter 10 for more on swarming). However, it takes a season or two of experience to judge when it's okay to bypass using a queen excluder (you can choose not to use the excluder if you are certain that there is a very strong nectar flow and the bees are rapidly filling many cells with nectar). My recommendation: Play it safe in year one and use the queen excluder. Next year you can try not using it when you put the honey supers on, timed to coincide with a major nectar flow.

## Shallow or medium honey super

Beekeepers use honey supers to collect surplus honey. That's *your* honey — the honey that you can harvest from your bees. You need to leave the honey in the deep-hive body for the bees. Supers are identical in design to the deep-hive bodies — and assemble in a similar manner — but the depth of the supers is shallower.

They come in two popular sizes: shallow and medium. The shallow supers are $5\frac{11}{16}$ inches deep, and the medium supers are $6\frac{5}{8}$ inches deep.

Honey supers are put on the hive about eight weeks after you first install your bees. For the second-year beekeeper, honey supers are placed on the hive when several different spring flowers start to bloom.

The reduced depth of the supers makes them easy to handle during the honey harvest. A shallow super with ten frames full of honey will weigh a hefty (but manageable) 35 to 40 pounds. A medium super with ten frames full of honey weighs about 50 to 55 pounds; however, a deep-hive body full of honey weighs a back-breaking 80 pounds or more. That's more weight than you'd want to deal with!

**TIP**

You can use medium-size equipment for your entire hive. Three medium-depth hive bodies are equivalent to two deep-hive bodies. Standardizing on one size means that all of your equipment is 100 percent interchangeable. The lighter weight of each medium-hive body makes lifting much, much easier. In addition, the eight-frame version of the Langstroth hive (versus the ten-frame) is considerably more manageable because of the lighter weight.

TIP

As the bees collect more honey, you can add more honey supers to the hive, stacking them one on top of another like so many stories to a skyscraper. For your first season, order one or two honey supers (either shallow or medium).

## Frames

Each wooden frame contains a single sheet of beeswax foundation (described in the next section). The frame is kind of like a picture frame. It firmly holds the wax and enables you to remove these panels of honeycomb for inspection or honey extraction. Ten deep frames are used in each deep-hive body (unless you have decided on using an eight-frame version of the Langstroth hive). Frames are the trickiest pieces of equipment you'll have to assemble. Beekeeping suppliers usually sell frames in packages of ten, with hardware included.

Although plastic frames and foundation are available from some beekeeping suppliers, I don't like the plastic products. I much prefer the all-natural equipment, and I feel the bees share my preference. So in this book, I focus on the traditional wooden frames and pure beeswax foundation.

There is no doubt that plastic won't rot, nor will it be nibbled up by critters. Plastic frames last longer than wood, and plastic foundation is far more durable than delicate wax foundation. However, the bees are very slow to work plastic foundation into honeycomb. You need a super strong nectar flow to entice them. Not so with the all-natural setup of wood frames and beeswax foundation. The bees will eagerly and quickly convert the beeswax foundation into honeycomb. And the natural stuff smells so good! Want to be convinced? Use plastic frames and plastic foundation in one hive and wooden frames and beeswax foundation in a second hive. See for yourself.

Frames come in three basic sizes: deep, shallow, and medium — corresponding to deep-hive bodies and shallow or medium honey supers. The method for assembling deep, shallow, or medium frames is identical. Regardless of its size, each frame has four basic components: one top bar with a wedge (the wedge holds the foundation in place), one bottom bar assembly (consisting of either two rails or a single bar with a slot running its length), and two sidebars (see Figure 5-3). Frames typically are supplied with the necessary and correct size nails.

TIP

Some suppliers offer a frame design that allows you to simply slide a sheet of foundation into a modified frame design, avoiding all the fuss associated with the more typical, standard frame design.

# MAKING YOUR WOODENWARE LAST

To get the maximum life from your equipment, you must protect it from the elements. If you don't, wood that's exposed to the weather rots.

After you've assembled your equipment and before you put your bees in their new home, paint all the outer surfaces of the hive (see the following list). Use (at least) two coats of a good quality outdoor paint (either latex or oil-based paints are okay). The color is up to you. Any light pastel color is fine, but avoid dark colors because they will make the hive too hot during summer. White seems to be the most traditional color. If you prefer, you can stain your woodenware and treat it with an outdoor grade of polyurethane.

If you have more than one hive, consider painting them different colors or adding some unique identifying marks at the entrance to help the bees visually distinguish which hive is their home. This helps prevent the bees from "drifting" into hives that are not theirs, and thus prevents the spreading of potential disease and parasites from one colony to the next.

**Do paint/treat the following:**

- Wooden hive-top feeder (outside surfaces only)
- Outer cover (inside and outside surfaces)
- Supers and hive bodies (outside surfaces only)
- Bottom board (all surfaces)
- Hive stand (all surfaces)

**Do not paint/treat the following:**

- Inner cover
- Frames (or top bars if you have a Top Bar hive)
- Inside surfaces of supers, hive bodies, and wooden hive-top feeder
- Queen excluder

URBAN

Sometimes in an urban area, it's best not to draw attention to the beehives. Painting them gray or some other drab color may camouflage them from prying eyes. Some urban beekeepers have painted their hives to resemble chimneys, brick walls, or even air-conditioning units. Clever!

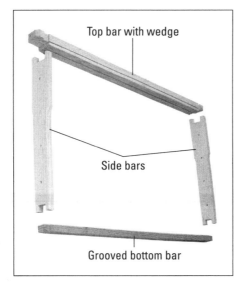

**FIGURE 5-3:**
The parts of a standard frame design.

Top bar with wedge

Side bars

Grooved bottom bar

*Courtesy of Howland Blackiston*

Assuming you are using the traditional design that is so readily available, assemble your frames by following these directions:

1. **Take the top bar and snap out the wedge strip.**

   You can use your hive tool to pry the wedge strip from its place. Clean up any *filigree* (rough edges) by scraping the wood with your hive tool. Save the wedge for use when you're installing the wax foundation (see the "Foundation" section next).

2. **Place the top bar on your tabletop work surface with the flat side facing down on the table.**

3. **Take the two side pieces and snap the wider end into the slots at either end of the top bar.**

4. **Now snap your bottom bar assembly into the slots at the narrow end of the side pieces.**

   Depending on the manufacturer, this assembly will either consist of two rails or a single bar with a slot running its length.

5. **Now nail all four pieces together.**

   Use a total of six nails per frame (two for each end of the top bar and one at each end of the bottom bar). In addition to nailing, I suggest you also glue the parts together using an all-weather wood glue. Doing so adds strength.

6. **Repeat these steps until all your frames are assembled.**

   Time for a break while the glue dries.

Don't be tempted to use any shortcuts. Frames undergo all kinds of abuse and stress, so their structural integrity is vital. Use glue for extra strength and don't skimp on the nails or settle for a bent nail that's partially driven home. If any nail comes through the wood to the outside, remove and replace it. There's no cheating when it comes to assembling frames!

If you're a visual person, check out my video on how to assemble frames at `www.dummies.com/go/beekeepingfd4e`.

# Foundation

Foundation consists of thin, rectangular sheets that are used to urge your bees to draw even and uniform honeycombs. It comes in two forms: plastic and beeswax. Using plastic foundation has some advantages because it's stronger than wax and resists wax moth infestations. But the bees are slow to accept plastic, and I don't recommend it for the new beekeeper. Instead, purchase foundation made from pure beeswax. The bees will accept it much more quickly than plastic, and you will have a much more productive and enjoyable first season with your bees. In subsequent seasons you can experiment with plastic — but I'll bet you come back to the wax! Beeswax foundation is imprinted with a hexagonal cell pattern on the wax that guides the bees as they draw out uniform, even combs. Some foundation comes with wire embedded into the foundation (my preference). Some you must wire manually after installing the foundation in the frames. That's a lot of extra work and takes a bit of skill. I suggest using the prewired foundation.

Your bees find the sweet smell of beeswax foundation irresistible and quickly draw out each sheet into thousands of beautiful, uniform cells on each side where they can store their food, raise brood, and collect honey for you!

To see a video of how to install beeswax foundation, go to `www.dummies.com/go/beekeepingfd4e`. The steps are the same as the ones outlined here.

Like frames, foundation comes in deep, shallow, and medium sizes — deep for the deep-hive bodies, shallow for the shallow supers, and medium for the medium supers. You insert the foundation into the frames the same way for all of them.

Here's how to insert foundation into your frames:

1. **With one hand, hold the frame upright on the table. Look closely at a sheet of foundation. If it's the prewired variety (my recommendation), you will note that vertical wires protrude from one side and are bent at right angles. However, wires at the other side are trimmed flush with the foundation. Drop this flush end into the long groove or slit of the bottom bar assembly and then coax the other end of the foundation into the space provided by your removal of the wedge bar (see Figure 5-4).**

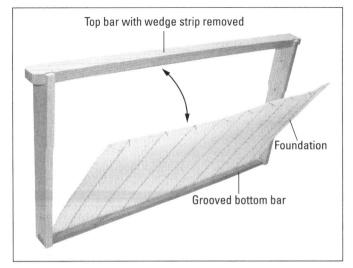

**FIGURE 5-4:** Inserting the prewired foundation sheet into the grooved or slotted bottom assembly.

Top bar with wedge strip removed

Foundation

Grooved bottom bar

*Courtesy of Howland Blackiston*

2. **Turn the frame and foundation upside down (with the top bar now resting flat on the table). Adjust the foundation laterally so equal space is on the left and right. Remember the wedge strip you removed when assembling the frames? Now's the time to use it! Return the wedge strip to its place, sandwiching the foundation's bent wires between the wedge strip and the top bar (see Figure 5-5). Use a brad driver to nail the wedge strip to the top bar (see Figure 5-6). Start with one 18-gauge brad in the center and then one brad at each end of the wedge strip. Add two more brads for good luck (five total).**

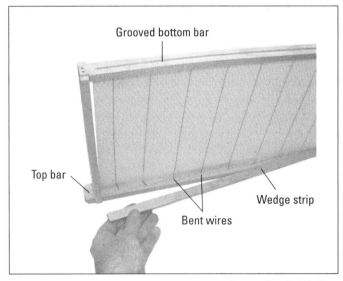

**FIGURE 5-5:** Turn the frame upside down to sandwich the foundation's bent wires between the wedge strip and the top bar.

Grooved bottom bar

Top bar

Wedge strip

Bent wires

*Courtesy of Howland Blackiston*

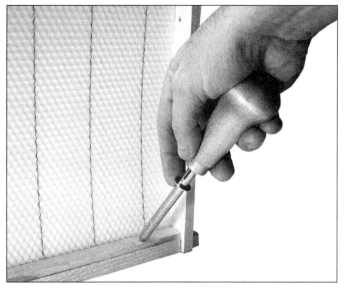

FIGURE 5-6:
With the frame still upside down, use a brad driver to nail the wedge strip back in place.

*Courtesy of Howland Blackiston*

3. **Finally, use support pins — they look like little metal clothespins — to hold the foundation securely in place (see Figure 5-7). The pins go through predrilled holes in the side bars and pinch the foundation to hold it in place. Although each side bar has three to four predrilled holes, use only *two* pins on each sidebar (four per frame). The extra holes are for those who want to manually wire their foundation — that's not me!**

FIGURE 5-7:
Insert support pins in predrilled holes of the side bars to hold the wax foundation in place.

*Courtesy of Howland Blackiston*

That's it! You've completed building one frame. There's a bunch more to go!

# Inner cover

Inner covers of good quality are made entirely of cypress or cedar wood. Budget models made from pressboard or Masonite also are available, but they don't seem to last as long. Alternatively, there are plastic ones available that will never rot. See Figure 5-8. The basic design consists of a framed flat plank with a precut hole in the center of the plank. The inner cover resembles a shallow tray (with a hole in the center). In some models, a notch is cut out of one of the lengths of frame. This is a ventilation notch, and it is positioned to the front of the hive. I prefer this design for the extra ventilation it provides. The inner cover is placed on the hive with the "tray" side facing up. See Figure 5-9. If your model has a half-moon ventilation notch (as seen in the figure) place the notch toward the front of the hive. The outer cover is placed over this inner cover.

**FIGURE 5-8:**
A wooden and a plastic inner cover.

*Courtesy of Howland Blackiston*

**FIGURE 5-9:**
The correct placement of an inner cover. Note tray side faces up.

*Courtesy of Howland Blackiston*

## Outer cover

Look for cypress or cedar wood when buying an outer cover. These woods resist rot and last the longest. Outer covers assemble in a manner similar to the inner cover: a frame containing flat planks of wood. But the outer cover typically has a galvanized steel tray that fits on the top, protecting it from the elements. Alternatively, there are some plastic models on the market that will never rot. Not quite as pretty as wood, but perhaps practical.

URBAN

Urban hives are often on rooftops with much higher winds to contend with. For the safety and comfort of your bees and those on the sidewalks below, use strong crank straps (ratchet tie-down straps) to secure all the hive parts together. One sudden gust of wind can send an outer cover flying off the rooftop and onto the street.

# Knowing the Basic Parts of a Top Bar Hive

A Top Bar hive is basically a long box with sides that slope at approximately a 30-degree angle. Viewed from the end, the box appears as a trapezoid, wide at the top and narrow on the bottom (see Figure 5-10). This is where a Top Bar hive departs from the dimensional standards of a Langstroth box.

Top Bar hives are not standard and different manufacturers make different sizes. The lengths of Top Bar hives vary but are generally between 36 and 48 inches long. The depth of the box can also vary, but most Top Bar hives measure about 10 inches in depth. Also, a Top Bar hive lends itself to home construction and with that comes your personal choice of size. See my book *Building Beehives For Dummies* (Wiley) for plans.

**FIGURE 5-10:**
The basic Top Bar
hive box. Note
the end is
trapezoidal.

*Courtesy of William Hesbach*

**TIP**

It's best to choose one size and stick with it for all your Top Bar hives. Just like when making the choice between eight- and ten-frame Langstroth equipment, you don't want to be stuck with bars and comb that are not interchangeable between your Top Bar hive colonies.

## The top bar

The top bar is a simple piece of ¾-inch wood that spans the width of the hive and is roughly 1⅜ inches wide. This is the part of the Top Bar hive where the bees attach the comb. In contrast, a Langstroth hive has frames and foundation where the bees attach the comb.

The top bar is usually equipped with a guide down the center that helps keep the comb straight as the bees build it (see Chapter 7). The center guides come in different designs based on the manufacturer, or you can choose your own style if you build your own top bars. The guides can be as simple as a groove with a piece of protruding wood, a small piece of foundation waxed into a groove, or a more elaborate cone shape. See Figure 5-11.

**TIP**

The guide on the top bar is the only part that supports the comb (which can get heavy when filled with honey). Pick a bar style that will allow the most surface area for the bees to attach the comb.

**FIGURE 5-11:**
The different types of center guides that help the bees get started building comb on the top bars.

*Courtesy of William Hesbach*

## Everything else

Top Bar hives also use outer covers, entrance reducers, and hive stands, all designed for this type of hive; their functions are similar to the same items for a Langstroth hive. Find out more about these components in the earlier "Knowing the Basic Woodenware Parts of the Langstroth Hive" section.

# Ordering Hive Parts

Hive manufacturers traditionally make their woodenware out of pine, cypress, or cedar. Hardwoods are fine, but too expensive for most hobbyists. A custom mahogany hive, for instance, runs more than $1,000 versus a standard pine and cypress hive for about $150 to $300. Many suppliers offer various grades of components from a *commercial* budget-grade to a *select* best quality-grade. Go for the highest quality that your budget allows. Although they may be a little more expensive upfront, quality parts assemble with greater ease and are more likely to outlast the budget versions.

Any of this stuff is available from beekeeping supply stores. Most of these vendors are now on the web. For a listing of some of my favorites, see Appendix A. And be sure to take advantage of the special discount coupons that are included at the end of this book.

**TIP**

Don't wait until the last minute to order and build your first startup kit. In the United States, springtime is the beginning of the beekeeping season. If you wait until spring to order your kit, you will likely have to wait to get it (the suppliers become swamped with orders at that time). Ideally it's best to get all the stuff you need and assemble your hives and frames a few months before you plan to install bees in your hive.

## Startup hive kits

Many suppliers offer a basic startup kit that takes the guesswork out of what you need to get. These kits often are priced to save a few bucks. Make certain your kit contains these basic items, discussed in this section:

>> Woodenware for the hive body/bodies

>> Honey super(s) if you are using a Langstroth hive

>> Frames and foundation if you are using a Langstroth hive, or top bars if you are using a Top Bar hive

>> Hardware to assemble stuff (various size nails and so on)

>> Veil and gloves

>> Smoker

>> Hive tool

>> Feeder (appropriate style for either Langstroth or Top Bar hive)

## Setting up shop

Before your bees arrive, you need to order and assemble the components that will become their new home. You don't need much space for putting the equipment together. A corner of the garage, basement, or even the kitchen will do just fine. A worktable is mighty handy, unless you like crawling around on the floor.

Get all your hive parts (woodenware) together and the instruction sheets that come with them. The only tool you *must* have is a hammer. But having the following also is mighty useful:

## ANTICIPATING THE LENGTH OF ASSEMBLY TIME

By all means make sure everything is ready way before your bees arrive on your doorstep. Don't wait until the last minute to put things together. It probably will take a bit longer than you think, particularly if you are doing this for the first time.

First-timers should allow three to four hours to assemble the various hive bodies and supers, bottom board, and the outer and inner covers. Assembling frames and installing the foundation may require another few hours. And then you need to allow an hour or two to paint your equipment. Plus there's the cleanup time and the time for the paint to dry. All in all, your weekend is cut out for you. Of course, you can order your kits preassembled by the vendor, but be prepared to pay for this service.

My advice? Order your equipment two to three months before your bees are scheduled to arrive. Use all that extra time to ensure a timely delivery of the equipment and to leisurely put things together long before the bees arrive.

>> A pair of pliers to remove nails that bend when you try to hammer them.

>> A brad driver with some ¾-inch, 18-gauge brads. Having this tool makes the installation of wax foundation go much faster.

>> A bottle of good-quality, all-weather wood glue. Gluing and nailing woodenware greatly improves their strength and longevity.

>> A carpenter's square to ensure parts won't wobble when assembled.

>> Some coarse sandpaper or a plane to tidy up any uneven spots.

>> A hive tool (I hope one came with your startup kit). It's pretty handy for pulling nails and prying off the frame's wedge strip.

**TIP**

Start assembling your equipment from the ground up. That means starting with the hive stand and moving on to all points north of that. That way you can begin to build the hive and ensure everything sits level and snug.

**WARNING**

If you are using a Langstroth hive, the various assembled components of the hive (bottom board, hive boxes, cover) are *not* nailed together. They simply are stacked one on top of the other (like a stack of pancakes). This enables you to open up and manipulate the hive and its parts during inspections.

# Adding on Feeders

Feeders are used to supply sugar syrup to your bees when the nectar flow is minimal or nonexistent. And for those who choose to do so, they also provide a convenient way to medicate your bees (some medications can be dissolved in sugar syrup and fed to your bees). New colonies need to be fed after installation. For established colonies, it's good management practice to feed your bees in the early spring and once again in autumn (see Chapter 9). Each of the many different kinds of feeders has its pluses and minuses. I've included a brief description of the more popular varieties.

**ALL NATURAL**

Feeding your colony can make a difference in its general well-being. In the spring, feeding sugar syrup stimulates the queen to lay more eggs, pushing colony expansion. Feeding in the fall encourages bees to store food and helps a colony survive the winter season.

## Hive-top feeder

The hive-top feeder (sometimes called a Miller feeder after its inventor, C. C. Miller) is the model I urge you to use for your Langstroth hive (see Figure 5-12). There are various models and variations on the original design, but the principle is similar from one to the next. As a new beekeeper, you will love how easy and safe it is to use. The hive-top feeder sits directly on top of the upper deep brood box and under the outer cover (no inner cover is used when a hive-top feeder is in place). It has a reservoir that can hold one to three gallons of syrup. Bees enter the feeder from below by means of a screened access.

**FIGURE 5-12:**
A hive-top feeder.

*Courtesy of Brushy Mountain*

The hive-top feeder has several distinct advantages over other types of feeders:

>> Its large capacity means you don't have to fill the feeder more than once every week or two.

>> The screened bee access means you can fill the feeder without risk of being stung (the bees are on the opposite side of the screen).

>> Because you don't have to completely open the hive to refill it, you don't disturb the colony (every time you smoke and open a hive, you set the bees' progress back a few days).

>> Because the syrup is not exposed to the sun, if you choose to medicate your bees, there is no light to diminish its effectiveness.

But with all of these good features, there are a couple of negatives:

>> Sometimes in the frenzy of feeding, bees will lose their grip on the screen, and some will drown in the syrup.

>> The hive needs to be level or nearly level for this feeder to be effective.

>> When it is full, the feeder is awkward and heavy to remove for routine inspections.

**URBAN**

The urban beekeeper is best off using a feeder that contains the most syrup because access to the hives is usually more difficult and therefore less frequent. Don't let your bees be without feed because their caretaker cannot scale the fire escape for a few days. Give them ample supplies of syrup and your rewards will be their loyalty and, ultimately, healthier and better honey producers. A hive-top feeder is a good choice for urban beekeepers.

## Entrance feeder

The entrance feeder (sometimes called a Boardman feeder) is a popular device for Langstroth hives. It consists of a small inverted jar of syrup that sits in a contraption at the entrance to the hive (see Figure 5-13). Entrance feeders are inexpensive and simple to use, and they come with many hive kits. But I don't recommend that you use an entrance feeder.

They have few advantages (other than being cheap) and have many worrisome disadvantages:

>> The feeder's proximity to the entrance can encourage bees from other hives to rob syrup and honey from your hive.

>> You're unable to medicate the syrup because it sits directly in the sun.

>> The feeder's exposure to the hot sun tends to spoil the syrup quickly.

>> Refilling the small jar frequently is necessary (often daily).

>> Using an entrance feeder in the spring isn't effective. The entrance feeder is at the bottom of the hive, but in the spring the bees cluster at the *top* of the hive.

>> Being at the entrance, you risk being stung by guard bees when you refill the feeder.

**FIGURE 5-13:** An entrance feeder.

*Courtesy of Howland Blackiston*

## Pail feeder

The pail feeder consists of a one-gallon plastic pail with a friction top closure. Several tiny holes are drilled in its top. The pail is filled with syrup, and the friction top is snapped into place. The pail then is inverted and placed over the center hole in the inner cover of a Langstroth hive (Figure 5-14). These work on the same principle as a water cooler where the liquid is held in by a vacuum. The syrup remains in the pail yet is available to the bees that feed by sticking their tongue through the small holes.

*Courtesy of Howland Blackiston*

Although inexpensive and relatively easy to use, it also has a few disadvantages:

>> This feeder is placed within an empty deep-hive body, with the outer cover on top.

>> You essentially must open the hive to refill the feeder, leaving you vulnerable to stings (see the following Tip to avoid this problem).

>> Refilling this feeder typically requires smoking your bees and disrupting the colony.

>> Its one-gallon capacity requires refilling once or twice a week.

>> Limited access to syrup means that only a few bees at a time can feed.

**TIP**

Cover the hole on the inner cover with a small piece of #8 wire hardware cloth; it will keep the bees from flying out at you when you remove the bucket for refilling. The hardware cloth should be affixed to the *top* side of the inner cover.

## Baggie feeder

Here's yet another cost-effective solution (see Figure 5-15). Pour 3 quarts of syrup into a one-gallon-size sealable plastic baggie. Zip it up. Lay the baggie of syrup flat and directly on the top bars. Note the air bubble that forms along the top of the bag. Use a razor blade to make a couple of 2-inch slits into the air bubble. Squeeze the bag slightly to allow some of the syrup to come through the slits (this helps the bees "discover" the syrup). Now you need to place an empty super and outer cover on the hive (to cover the feeder). This type of feeder works best on a Langstroth hive.

Courtesy of Howland Blackiston

**FIGURE 5-15:** A baggie feeder is a cost-effective feeding option.

The advantages of using a baggie feeder are as follows:

>> Very cost effective

>> Reduces the likelihood of robbing

>> Puts the feed directly on top of the bees for easy access

>> No drowned bees

There are some disadvantages:

>> You have to open the hive and disrupt the bees to put new bags on.

>> The old bags are not reusable after you cut them with a razor.

>> The bags have to be replaced fairly frequently.

## Frame feeder

This plastic feeder (often called a Boardman feeder) is a narrow vessel resembling a standard frame that is placed within the upper Langstroth hive body, replacing one of the frames next to the outside wall (see Figure 5-16). Filled with a pint or two of syrup, bees have direct access to it. But it isn't very practical for the following reasons:

**FIGURE 5-16:** Frame feeders are placed within the hive, replacing a frame of comb.

>> Its capacity is small so it must be refilled frequently, sometimes daily.

>> You lose the use of one frame while the feeder is in place.

>> Opening the hive to refill the feeder is disruptive to the colony and exposes you to stings.

>> Bees can drown in the feeder.

## Top Bar hive feeders

If you decide on a Top Bar hive, it will most likely come with plans to make a feeder or come with one you can just fill and use. Otherwise, Top Bar hive feeders fall into the category of in-hive feeders such as the baggie feeder, which is placed on the bottom of the box, not on the top bars as in the Langstroth hive, or the frame feeder mentioned previously. Other options include jars that fit inside trapezoidal-shaped containers, or simple plastic containers placed inside the hive and filled with syrup. See Figure 5-17.

*Courtesy of William Hesbach*

Because there isn't a booming commercial market for Top Bar hives, most Top Bar feeders wind up being homemade. You can find simple instructions on the Internet.

# Fundamental Tools

Two tools — the smoker and the hive tool — are a musts for the beekeeper. They're used every time you visit any kind of hive and are indispensable.

## Smoker

The smoker will become your best friend. Smoke changes the bees' behavior and enables you to safely inspect your hive. Quite simply, the smoker is a fire chamber with bellows designed to produce lots of cool smoke. Figure out how to light it so it stays lit, and never overdo the smoking process. A little smoke goes a long way. (See Chapter 7 for more about how to use your smoker.)

**TIP**

Be sure to go online and watch my videos on how to light and properly use a smoker. Just go to www.dummies.com/go/beekeepingfd4e.

Smokers come in all shapes, sizes, and price ranges, as shown in Figure 5-18. The style that you choose doesn't really matter. The key to a good smoker is the quality of the bellows. Consider one fabricated from stainless steel to avoid rusting.

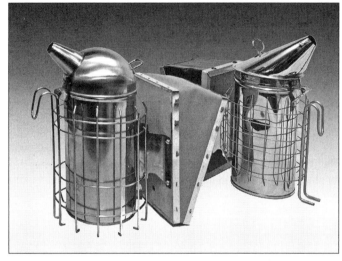

*Courtesy of Howland Blackiston*

**FIGURE 5-18:** Various smokers.

## Hive tool and frame lifter

The versatility of the simple hive tool is impressive. Don't visit your hives without one! Use it to scrape wax and propolis off woodenware. Use it to loosen hive parts, open the hive, and manipulate frames. In addition, a frame lifter is used by many beekeepers to firmly grip a frame and remove the frame from the hive. Bee suppliers offer a wide range of styles and designs to choose from. The ones pictured in Figure 5-19 are among the most commonly used.

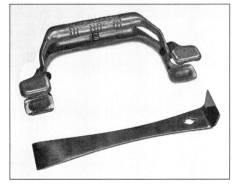

**FIGURE 5-19:** A frame lifter (top) and a hive tool (bottom).

*Courtesy of Howland Blackiston*

# Bee-Proof Clothing

At minimum, new beekeepers should wear a long-sleeved shirt when visiting the hive. Light colors are best — bees don't like dark colors (black, navy, and so on). Wear long pants and slip-on boots. Tuck your pant legs into the boots. Alternatively, use Velcro or elastic strips (even rubber bands) to secure your pant legs closed. You don't want a curious bee exploring up your leg! Better yet, order a bee suit (coveralls) or a bee jacket from your beekeeping supplier. You should also invest in veils and gloves, which are discussed in this section.

## Veils

Don't ever visit your hive without wearing a veil. Although your new colony of bees is likely to be super gentle (especially during the first few weeks of the season), it defies common sense to put yourself at risk. As the colony grows and matures, you will be working with and among upwards of 60,000 bees.

**TIP**

It's not that the bees are aggressive (they're not), but they are super curious. They love to explore dark holes (like your ear canal and nostrils). Don't tempt fate — wear a veil.

Veils come in many different models (see Figure 5-20) and price ranges. Some are simple veils that slip over your head; others are integral to a pullover jacket or even a full coverall. Pick the style that appeals most to you. If your colony tends to be more aggressive, more protection is advised. But remember, the more that you wear, the hotter you'll be during summer inspections.

**URBAN**

Urban beekeepers may have more to lose by not wearing a veil. Whereas the suburban beekeeper might suffer a sting and run away from the hostility, the city-dwelling beekeeper risks a much more devastating fate if a bee enters her nose or ear — one loses all sense of space and equilibrium when a bee stings these areas, and if the hives are located on a roof, the sting is the least of the beekeeper's problems! You certainly don't want to flee and wind up falling off the roof! Roof-top beekeepers should always use a veil.

**TIP**

Keep an extra veil or two on hand for visitors who want to watch while you inspect your bees.

## Gloves

New beekeepers understandably like the idea of using gloves (see Figure 5-21), but I urge you not to use them for installing your bees or for routine inspections.

You don't really need them at those times, especially with a new colony or early in the season. Gloves only make you clumsier. They inhibit your sense of touch, which can result in your inadvertently injuring bees. That's counterproductive and only makes them more defensive when they see you coming.

FIGURE 5-20: Protective clothing comes in various styles, from minimal to full coverage. This beekeeper uses a veil-and-jacket combination, leather gloves, and high boots to keep him bee-tight.

*Courtesy of Howland Blackiston*

The only times that you really need to use gloves are

>> Late in the season (when your colony is at its strongest)

>> During honey harvest season (when your bees are protective of their honey)

>> When moving hive bodies (when you have a great deal of heavy work to do in a short period of time)

Other times leave the gloves at home. If you must, you can use heavy gardening gloves, or special beekeeping gloves with long sleeves (available from beekeeping supply vendors).

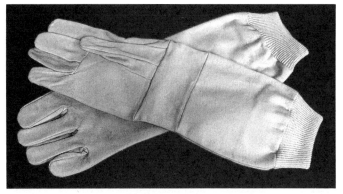

**FIGURE 5-21:**
Although not
needed for
routine inspec-
tions, it's a good
idea to have a
pair of protective
gloves. Maybe
even an extra pair
for visitors.

*Courtesy of Howland Blackiston*

# Really Helpful Accessories

All kinds of gadgets, gizmos, and doodads are available to the beekeeper. Some are more useful than others. I describe a few of my favorites in this section.

## Elevated hive stand

I have all my hives on elevated stands. Elevated hive stands are something you're more likely to build than purchase. I've included detailed plans for making an elevated hive stand in Chapter 18. Alternatively, you can fashion an elevated stand from a few cinderblocks (see Figure 5-22). You can also use posts of various sorts (see Figure 5-23), or purchase a commercial-grade stand that's waterproof and comes with a built-in frame rest (see Figure 5-24). In any case, having the hive off the ground means less bending over during inspections and makes the hive far easier to work with.

In the case of Top Bar hives, an elevated stand can be attached directly to the body to make the hive a comfortable working height (see Figure 5-25) or fabricated to cradle the body of the hive and be more portable. Any stand that elevates the Top Bar hive to about 36 inches (typical countertop height) will work just fine.

**TIP**

To help prevent ants from nesting under your top cover, the legs of the stand can rest in a container of liquid large enough to act as a moat and prevent ants from crawling up the legs. Raising the entrance using an elevated hive stand also is a natural way to deter skunks from snacking on your bees. (See Chapter 13 for more information on heading off ants, skunks, and other honey-bee pests.)

**FIGURE 5-22:**
You can elevate
your hive on
cinderblocks.

*Courtesy of Howland Blackiston*

**FIGURE 5-23:**
You can use a
level stump to get
your hive up off
the ground.

*Courtesy of Howland Blackiston*

**FIGURE 5-24:**
This commercially
available
hive stand is
completely
weatherproof
and includes
a built-in
frame rest.

*Courtesy of Beesmart.com*

**FIGURE 5-25:**
These hive stands
elevate a Top
Bar hive to a
comfortable
working height
and make
inspections and
harvesting less
physically
stressful for you.

*Courtesy of William Hesbach*

## Frame rest

For users of a Langstroth hive, a frame rest is a super-helpful device that I love. This gadget hangs on the side of the hive, providing a convenient and secure place to rest frames during routine inspections (see Figure 5-26). It holds up to three frames, giving you plenty of room in the hive to manipulate other frames without crushing bees.

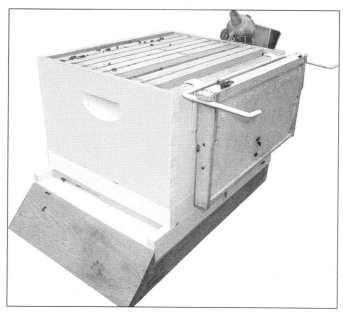

*Courtesy of Howland Blackiston*

## Bee brush

The long, super-soft bristles of a bee brush enable you to remove bees from frames and clothing without hurting them (see Figure 5-27). Some beekeepers use a goose feather for this purpose. Keep that in mind in the event you have an extra goose around the house.

*Courtesy of Howland Blackiston*

## Slatted rack

You may want to sandwich a slatted rack between the Langstroth hive's bottom board and lower deep-hive body (see Figure 5-28). It does an excellent job of helping air circulation throughout the hive. Also, no cold drafts reach the front of the hive, which in turn encourages the queen to lay eggs right to the front of the combs. More eggs mean more bees, stronger hives, and more honey for you! I use a slatted rack on all my hives.

**FIGURE 5-28:**
Slatted racks
help improve
ventilation and
can improve
brood pattern.

*Courtesy of Howland Blackiston*

## Screened bottom board

The screened bottom board replaces the standard bottom board. Its bottom is completely open, except for the #8 hardware cloth screening that makes up its floor. The unit also has a removable "sticky board" that you can use to monitor the colony's varroa mite population. These products have become a standard part of integrated pest management (IPM) and are also available for Top Bar hives (see Figure 5-29).

**ALL
NATURAL**

With Varroa mites a problem for many beekeepers, screened bottom boards are gaining popularity. (See Chapter 13 for more information on Varroa mites.) A moderate percentage of mites naturally fall off the bees each day and land on the bottom board of the hive. Ordinarily they just wait to reattach themselves to the bees. But when you use a screened bottom board in place of a regular bottom board, mites fall to the ground, or they are trapped on a "sticky board" placed *under* the screening. Either way, they are unable to crawl back up into the hive. When using a sticky board, the beekeeper can actually count the number of mites that have fallen off the bees and thus monitor the mite population.

There is another great advantage of using a screened bottom board — improved ventilation. Poor ventilation is one of the leading causes of stress on the colony. Using a screened bottom board (sans sticky board) provides the ultimate in ventilation.

I am often asked if you use a slatted rack with a screened bottom board. The answer is no. Use one or the other. Either one will improve ventilation, but only the screened bottom board will help you monitor the mite population.

**FIGURE 5-29:**
Screen bottom
boards are
available for Top
Bar hives.

*Courtesy of William Hesbach*

# Beekeeper's toolbox

It seems that whenever I get out in the yard and start an inspection, there's something I need that I don't have with me. I've pretty much solved that annoying situation by bringing along a toolbox with everything I might possibly need during an inspection. Here's a list of some of the goodies in my beekeeping toolbox (see Figure 5-30):

» **Basic beekeeping tools:** A hive tool and frame lifter are absolute necessities to help you get into the hive.

» **Matches:** You'll need these to relight your smoker. Keep some strike-anywhere matches in a waterproof container.

» **A spray bottle of alcohol:** Fill a small plastic spray bottle with plain rubbing alcohol. Use this during inspections to clean any sticky honey or pollen off your hands. Never spray the bees with this!

» **Baby powder:** Dust your hands with baby powder before inspections. The bees seem to like the smell, and it helps keep your hands clean.

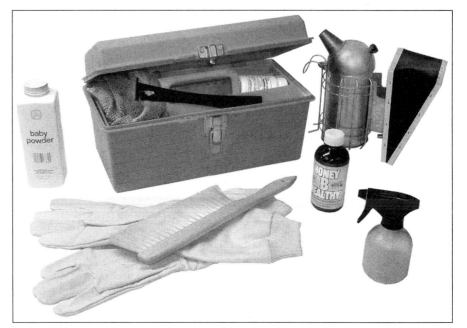

**FIGURE 5-30:**
This simple toolbox (formerly a fishing tackle box) is a convenient way to tote basic supplies to the bee yard.

*Courtesy of Howland Blackiston*

» **Disposable latex gloves:** Available at any pharmacy, I use these during inspections late in the season when propolis is plentiful. The gloves don't impede my dexterity, and they keep my hands clean when working among the authoritatively sticky propolis. ***Note:*** They are not sting-proof.

» **Notebook and pencil:** You can take notes on your inspections as well as note what tasks you have completed and what may need attention on a subsequent visit.

» **Hammer and nails:** A hammer and an assortment of frame and hive nails are helpful for any quick onsite repairs that need attention.

» **Multipurpose jackknife:** You never know when you might need those little pliers, screw drivers, or blades.

» **Reading glasses and flashlight:** Tiny eggs can be hard to see without a little help.

» **A long kitchen knife:** This is handy for Top Bar hive beekeepers to cut bridge comb that tends to "glue" top bars together.

**URBAN**

When doing inspections on an urban rooftop hive, bring along a heavy sheet, a light blanket, or a large piece of cardboard. The rooftops can get much hotter than the air, and when bees are knocked from frames during inspections, it is better to have them land on a relatively cool blanket than to have them scorch their six little feet on the hot tar. In the heat of the summer, this is the difference between them surviving or not surviving a routine inspection.

# Chapter **6**

# Obtaining and Installing Your Bees

Ordering bees and putting them into their new home (hiving) is just about my favorite part of beekeeping. *Hiving* bees is surprisingly easy — and a lot safer than you might imagine. You don't often get an opportunity to do it because once your bees are established, you don't typically need to purchase a new colony. Bees are perennial and remain in their hive generation after generation. Only when you start a new hive or lose a hive to disease or starvation do you need to buy and install a new colony of bees.

I was a nervous wreck in the days and hours before installing my first colony. Like an expectant father, I paced the floor nervously until the day they arrived. And when they arrived, I fretted about how in the world I'd get all those bees into the hive. Would they fly away? Would they attack and sting me? Would the queen be okay? Would I do the right things? Help! All my fears and apprehensions turned out to be unfounded. It was as easy as pie and a thoroughly delightful experience.

## Determining the Kind of Bee You Want

Across the globe there are numerous kinds of honey bees. You can choose from many different strains and hybrids of honey bees. Each has its own pluses and minuses. The following list acquaints you with some of the more common types of bees. Most of these types are readily available from bee suppliers. Some suppliers even specialize in particular breeds, so shop around to find what you want.

- » **Italian** *(A. m. ligustica)*: Hands down, the most popular bee in North America and Europe. These honey bees are a pretty yellow-brown in color with distinct dark bands. This race originally hails from the Apennine Peninsula in Italy. They are calm bees and good comb producers, and the large brood that Italian bees produce results in quick colony growth. They maintain a big winter colony, however, which requires large stores of food. You can help offset this by feeding them before the onset of winter (see Chapter 9).

- » **Carniolan** *(A. m. carnica)*: Currently these are the second-most popular bee in North America. These bees are dark in color with broad gray bands. They originally hail from the mountains of Austria and Slovenia. They overwinter well and tend to have fewer health issues than some other races. But this type also exhibits a strong tendency to swarm. Carniolans maintain a small winter colony, which requires smaller stores of food.

- » **Caucasian** *(A. m. caucasica)*: Caucasian bees are mostly gray in color and are extremely adaptable to harsh weather conditions. They hail from the Caucasian Mountains near the Black Sea. They are gentle and calm. Caucasians make extensive use of propolis to chink up drafty openings, which can make quite a sticky challenge for the beekeeper. These bees also are prone to robbing honey, which can create a rather chaotic bee yard. They can also fall victim more easily to Nosema disease, so be sure to take preventive steps for your Caucasian bees (see Chapter 12). For these reasons, this type of bee has lost some favor among backyard beekeepers in recent years.

- » **Russian:** In the 1990s, efforts to find a honey bee that was resistant to Varroa and tracheal mites led United States Department of Agriculture (USDA) researchers to Russia, where a strain of honey bee seemed to have developed a resistance to the pesky mites. Indeed, the Russian bees seem to be better at coping with the parasites that have created so much trouble for other strains of bees. These bees have tend to curtail brood production when pollen and nectar are in short supply, resulting in a smaller winter colony — a helpful trait that leads to better success when it comes to overwintering in cold climates. I've had good success with Russian bees. Since 2000, Russian queen bees have been available from some bee breeders. They are worth considering.

- » **Buckfast (hybrid):** The Buckfast bee was the creation of Brother Adam, a Benedictine monk at Buckfast Abby in the United Kingdom. Brother Adam earned a well-deserved reputation as one of the most knowledgeable bee breeders in the world. The precise heritage of the Buckfast bee seems to have been known only by Brother Adam. He mixed the British bee with scores of bees from other races, seeking the perfect blend of gentleness, productivity, and disease resistance. The Buckfast bee's resulting characteristics have created quite a fan club of beekeepers from all around the world.

The Buckfast bee excels at brood rearing but exhibits a tendency toward robbing and absconding from the hive (see Chapter 10 for information on how to prevent these bad habits). Buckfast bees are harder to find than the others mentioned, and even if you do find a breeder selling them, the authenticity of their genetics (as compared to Brother Adam's strain) may be suspect. These days, some breeders in Canada offer Buckfast bees.

>> **Cordovan (hybrid):** Here's a gentle bee that's gaining some interest. This race is often used as a marker in research because they are so easy to distinguish from other strains — they are a brilliant bright yellow, particularly the queen, who is even more yellow than the rest of the colony. This makes the queen easier to find, and some beekeepers love the convenience of being able to quickly spot the queen. The downside is that Cordovans are not particularly good honey producers (which is not an issue for researchers and those just wanting bees for pollination). In addition, they don't do very well in cold, wet climates. But they certainly are pretty. You can find some breeders of this race in California and the Southeastern United States.

**WARNING**

You may note ads for bees touted as being "hygienic." *Hygienic* bees are those that are more likely to naturally detect and remove Varroa mites and slow the development of a Varroa mite population in a hive. Such behavior is admirable and helps ensure better overall colony health. But beware! Not all who advertise hygienic bees are actually doing testing to know if they are, in fact, breeding bees for this desired behavior. If in doubt, query the individual as to what specifically he is doing and how he knows he has more hygienic stock. More than a little exaggeration takes place when it comes to marketing packages of honey bees.

>> **Africanized (hybrid):** The list of bee races is not complete without a nod to the so-called Killer Bee. This bee is not commercially available, nor desirable to have. I mention it here because its presence has become a reality throughout South America, Mexico, and much of the southern United States. This bee's defensive behavior makes it difficult and even dangerous to manage. (See Chapter 10 for more on this type of bee.)

**TIP**

Another point to consider is raising your own survivor queens that exhibit the kinds of behaviors and characteristics you want for your situation. Many experienced beekeepers are using this approach to naturally improve the overall health and demeanor of their colonies. Learn more about queen rearing in Chapter 14.

Generally speaking, the four characteristics that you should consider when picking out the bee strain that you want to raise are gentleness, productivity, disease tolerance, and how well the bees tolerate the weather conditions in which you will be raising your girls. Table 6-1 assigns the various types of bees previously mentioned a rating from 1 to 3 in these four categories, with 1 being the most desirable and 3 the least desirable.

**TABLE 6-1**   Characteristics of Various Common Honey Bee Types

| Bee Type | Gentleness | Productivity | Disease Tolerance | Wintering in Cold Climates |
|---|---|---|---|---|
| Italian | 1 | 1 | 2 | 2 |
| Russian | 1 | 1 | 1 | 1 |
| Carniolan | 1 | 2 | 2 | 2 |
| Caucasian | 1 | 2 | 2 | 1 |
| Buckfast (hybrid) | 2 | 2 | 1 | 1 |
| Cordovan (hybrid) | 1 | 3 | 2 | 3 |
| Africanized | 3 | 1 | 1 | 3 |

TIP

After all that's said and done, which kind of bee do I recommend you start with? Try the Italian, Carniolan, or Russian. No doubt about it. They are gentle, productive, and do well in many different climates. These are great bees for beginning beekeepers. Look no further than these three in your first year.

At some point in years to come, you may want to try raising your own various races and hybrids of bees. Much is involved in breeding bees. It's a science that involves a good knowledge of biology, entomology, and genetics. A good way to get your feet wet is to try raising your own queens, which allows you to retain the desirable characteristics of your favorite colonies. For a primer in raising queens, see Chapter 14.

URBAN

When beekeeping in tight quarters where neighbors are in close proximity, it's prudent to select gentle bees that are not prone to frequent swarming.

# Deciding How to Obtain Your Initial Bee Colony

You'll need some bees if you're going to be a beekeeper. But where do they come from? You have several options when it comes to obtaining your bees. Some are good; others are not so good. This section describes these options and their benefits or drawbacks.

## Ordering package bees

One of your best options and by far the most popular way to start a new Langstroth or Top Bar hive is to order package bees. It's the choice that I most recommend.

You can order bees by the pound from a reputable supplier. In the United States, *bee breeders* are found mostly in the southern states. Most will ship just about anywhere in the continental United States.

A package of bees and a single queen are contained in a small wooden or plastic box with two screened or ventilated sides (see Figure 6-1). Packaged bees are either shipped directly to the beekeeper or the beekeeper picks up the package from the supplier. A package of bees is about the size of a large shoebox and includes a small, screened cage for the queen (about the size of a matchbook) and a tin can of sugar syrup that serves to feed the bees during their journey. A three-pound package of bees contains about 10,000 bees, the ideal size for you to order. Order one package of bees for each hive you plan to start.

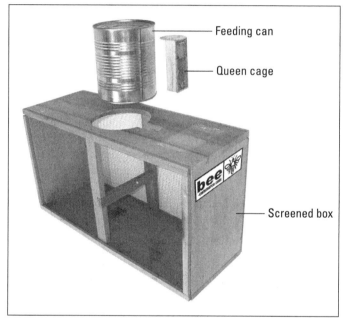

**FIGURE 6-1:**
Package bees are contained in either wooden or plastic boxes. Note the feeding can and the queen cage.

*Courtesy of Howland Blackiston*

TIP

Order a marked queen with the package. *Marked* means that a small colored dot has been painted on her thorax. Vendors charge a little more for marking, but this dot helps you spot the queen in your hive during inspections. It also confirms that the queen you see is the one that you installed (versus discovering an unmarked one that means your queen is gone and another has taken her place). The color of the dot indicates the year your queen was purchased (a useful thing to know because it allows you to keep track of the queen's age — you will want to replace her every couple of years to keep brood production optimized). Chapter 14 includes the international color code for marking queens.

Be sure to pick a reputable dealer with a good track record for providing healthy and disease-free package bees (criteria for selecting a vendor is discussed in "Picking a Reputable Bee Supplier" later in this chapter). When ordering, be sure to ask to see a copy of a certificate of health from the vendor's state apiary inspector. If the vendor refuses, be wary.

City dwellers can have a challenge getting package bees. Often they don't have cars to drive to bee breeders, and it may not be practical to get bees shipped to an apartment building. Check with urban beekeeping clubs; they will sometimes band together to arrange for their members to pick up from a truckload of package bees.

## Buying a "nuc" colony

For Langstroth beekeepers, buying a nuc is another good option for the new beekeeper: Find a local beekeeper who can sell you a *nucleus (nuc)* colony of bees. A nuc consists of four to five frames of brood and bees, plus an actively laying queen. All you do is transfer the frames (bees and all) from the nuc box into your own hive. The box usually goes back to the supplier. But finding someone who sells nucs isn't necessarily so easy because only a few beekeepers produce nucs for sale. After all, raising volumes of nucs for sale is a whole lot of work. But if you *can* find a local source, it's far less stressful for the bees (they don't have to be shipped). You can also be reasonably sure that the bees will do well in your geographic area. After all, it's already the place they call home! An added plus is that having a local supplier gives you a convenient place to go when you have beekeeping questions (your own neighborhood bee mentor). To find a supplier in your neck of the woods, do an online search for "beekeeping" and your zip code, city, or state; call your state's bee inspector; or ask members of a local beekeeping club or association.

For Top Bar hive beekeepers, purchasing a nucleus colony is usually not an option because of the nonstandard size of hive boxes. But, one way you may find a nuc for your Top Bar hive is to contact someone in your local bee club or join a TBH Facebook page and post a request — it's a great way to start valuable connections.

To find a bee club or association in your state, hop on the Internet and go to www.beeculture.com. Click the "Resources" tab, select "Find a Local Beekeeper," and then choose your state. You will find a listing of all the bee clubs and associations in your area.

A *nuc* or *nucleus* consists of a small wooden or cardboard hive (a "nuc box"; see Figure 6-2) with four to five frames of brood and bees, plus a young queen.

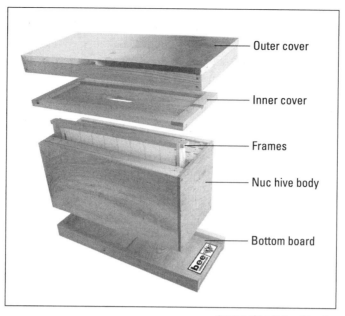

**FIGURE 6-2:**
A standard
nuc box.

Outer cover

Inner cover

Frames

Nuc hive body

Bottom board

*Courtesy of* Bee Culture Magazine

Look for a reputable dealer with a good track record for providing healthy bees (free of disease). Ask whether the state bee inspector inspects the vendor annually. Request a copy of a certificate of health from the state. If you can find a reputable beekeeper with nucs for sale, this is a convenient way to start a hive and quickly build up a strong colony.

## Purchasing an established colony

You may find a local beekeeper who's willing to sell you a fully established colony of bees — hive, bees, the whole works! This is fine and dandy, but more challenging than I recommend for a new beekeeper. First, you encounter many more bees to deal with than when you just get a package or nuc. And the bees are mature and well established in their hive. They tend to be more protective of their hive than a newly established colony (you're more likely to get stung). Their sheer volume makes inspecting the hive a challenge for someone just getting started in beekeeping. Furthermore, old equipment may be harder to manipulate (things tend to get glued together with propolis after the first season). More important, you also lose the opportunity to discover some of the subtleties of beekeeping that you can experience only when starting a hive from scratch: the building of new comb, introducing a new queen, and witnessing the development of a new colony.

Wait until you've gained more experience as a beekeeper before you purchase an established colony. If you're determined, however, to select this option, make sure you have your state's apiary inspector or an experienced local beekeeper look at the colony before you agree to buy it. You want to be 100 percent certain the colony is free of disease (for more information about honey bee diseases, see Chapter 12). After all, you wouldn't buy a used car without having a mechanic look at it first.

**URBAN**

It likely goes without saying, but if you are an urban beekeeper who plans to keep your bees on a rooftop, lugging an established, full-blown colony up a flight of stairs and onto a roof is not a prudent choice.

## Capturing a wild swarm of bees

Here's an option where the price is right: Swarms are free. But I don't recommend this for the first-year beekeeper. Capturing a wild swarm is a bit tricky for someone who has never handled bees. And you can never be sure of the health, genetics, and temperament of a wild swarm. In some areas (mostly the deep southern United States) you face the possibility that the swarm you attempt to capture may be *Africanized* (see Chapter 10). My advice? Save this adventure for year two, and watch an experienced beekeeper capture a swarm before trying it yourself.

**WARNING**

You may be tempted to get your first bee colony by transferring feral (wild) bees from a tree hollow or from the walls of a building. After all, they are, like a swarm, free. Resist the temptation! Any number of things can go wrong in moving bees from an established nest site. Leave wild bees alone until you have more experience. And even then, first see it done by an expert. Then decide if that's something you ever want to try!

**TIP**

You can find information about capturing a swarm in Chapter 10.

# Picking a Reputable Bee Supplier

By checking advertisements in bee journals and surfing the Internet you'll come up with a long list of bee suppliers (see Appendix A for a list of my favorite suppliers along with information on bee-related websites, journals, and organizations). But all vendors are not created equal. Here are some rules for picking a good vendor:

>> **Be sure to pick a well-established vendor who has been selling bees for many years.** The beekeeping business is full of well-meaning amateurs who get in and out of supplying bees. They lack experience, which can result in poor quality bees and lackluster customer service.

>> **Look for a vendor with a reputation for consistently supplying healthy bees and providing dependable shipping and good customer service.** Figure 6-3 shows a picture of a well-run commercial bee-breeding yard.

**FIGURE 6-3:**
The yard of a commercial bee supplier.

*Courtesy of* Bee Culture Magazine

>> **Ask whether the original supplier of the bees is inspected each year by the state's apiary inspector.** Request a copy of its health certificate. If the supplier refuses to comply, look elsewhere.

>> **Ask potential suppliers about their replacement guarantee.** A reputable supplier replaces a package of bees that dies during shipment.

>> **Be suspicious of suppliers who make extravagant claims.** Some walk a fine ethical line when they advertise that their bees are "mite or disease resistant." No such breed of bee exists. New beekeepers are easy prey for these charlatans. If the claims sound too good to be true, they probably are. Look elsewhere.

>> **Consult with representatives of regional bee associations.** Contact your state's apiary inspector or other bee association representatives. Find out whom they recommend as suppliers. Get them to share their experiences with you — good and bad.

## DECIDING HOW MANY HIVES YOU WANT

Starting your adventure with *two* hives of bees offers certain advantages. Having two gives you a basis for comparison. It enables you to borrow frames from a stronger colony to supplement a weaker colony. In some ways two hives double the fun. You'll have more bees to pollinate your garden and more opportunities to witness what goes on within a colony. And, of course, you'll double your honey harvest! You can also double the rate of your learning curve. I suggest, however, that you begin with no more than two hives during your first year. More than two can be too much for the beginner to handle. Too many bees can be too time consuming and present too many new problems to digest before you really know the subtleties of beekeeping.

>> **Join a local bee club to get vendor recommendations from other members.** This also is a great way to find out more about beekeeping and latch onto a mentor. Many clubs have "new beekeeper" programs and workshops.

# Deciding When to Place Your Order

When you're ordering packaged bees, you want to time your order so you receive your bees as early in the spring as the weather allows. Doing so gives your colony time to build its numbers for the summer "honey flow" and means your bees are available for early pollination. Suppliers usually start shipping packaged bees early in April and continue through the end of May. Large commercial bee breeders shake bees into screened packages and ship hundreds of packages daily during this season (see Figure 6-4). After that, the weather simply is too hot for shipping packaged bees — they won't survive the trip during the scorching hot days of summer (most bees ship from the southern states). Local bee suppliers have nucs available in a similar time frame.

**TIP**

Don't wait until springtime to order your bees. Bees are in limited supply and available on a *first-ordered, first-shipped basis.* Avoid disappointment. Place your order very early. Ordering in October or November for delivery the following spring is not too early!

FIGURE 6-4:
A commercial bee breeder shakes bees into packages for shipping to beekeepers.

*Courtesy of* Bee Culture Magazine

# The Day Your Girls Arrive

Unless you're picking up your bees directly from the breeder, you may not know the exact day that your bees will arrive at your doorstep. But many suppliers will at least let you know the approximate week they plan to *ship* your package bees. Weather (rain, temperature) is a determining factor for the bee suppliers, so be patient as they wait for the weather to make it safe for shipping.

**TIP**

If your package of bees is being mailed to you, about a week before the anticipated date of arrival, alert your local post office that you're expecting bees. Make sure you provide the shipper with your telephone number so you can be reached the moment your bees come in. If your bees are arriving by mail, in most communities the post office asks that you pick up your bees at the post office. Instruct the post office that the package needs to be kept in a cool, dark place until you arrive.

**WARNING**

Bees arriving at your door are *not* the signal for you to start assembling your equipment. Plan ahead! Make sure everything is ready for your girls before they arrive.

## Bringing home your bees

When the bees finally arrive, follow these steps in the order they are given:

1. **Inspect the package closely.**

   Make sure your bees are alive. You may find some dead bees on the bottom of the package, but that is to be expected. If you find a full inch or more of dead bees on the bottom of the package, however, fill out a form from your shipper and call your vendor. He or she should replace your bees if most or all the bees are dead.

2. **Take your bees home right away (but don't put them in the hot, stuffy trunk of your car).**

   They'll be hot, tired, and thirsty from traveling.

3. **When you get home, spray the package liberally with cool water using a clean mister or spray bottle.**

4. **Place the package of bees in a cool place, such as your basement or garage, for an hour.**

5. **After the hour has passed, spray the package of bees with sugar syrup (see the recipe that follows).**

   Don't *brush* syrup on the screen because doing so literally brushes off many little bee feet in the process.

TIP

You must have a means for feeding your bees once they're in the hive. I strongly recommend using a good-quality hive-top feeder. Alternatively you can use a feeding pail or a baggie feeder (see Chapter 5 for additional information on different kinds of feeders).

## Recipe for sugar syrup

You'll likely need to feed your bees sugar syrup twice a year (in spring and in autumn).

The early spring feeding stimulates activity in the hive and gets your colony up and running fast. It also may save lives if the bees' stores of honey have dropped dangerously low.

The colony will store the autumn sugar syrup feeding for use during the cold winter months (assuming your winter has cold months).

In either case, feeding syrup is also a convenient way to administer medications if you decide to use a medicated approach to bee health. (More on that in Chapters 12 and 13.)

For springtime syrup, boil 5 pints (2½ quarts) of water on the stove. When it comes to a rolling boil, turn off the heat and add 5 pounds of white granulated sugar. Be sure you turn off the stove. If you continue boiling the sugar, it may caramelize, and that makes the bees sick. Stir until the sugar completely dissolves. The syrup must cool to room temperature before you can feed it to your bees.

TIP

Here's a way to remember this syrup ratio: Mix a pound (16 ounces) of sugar with a pint (16 ounces) of water. And recall the old saying that "a pint's a pound, the world around."

# Putting Your Bees into the Hive

The fun stuff comes next. Sure, you'll be nervous, but that's only because you're about to do something you've never done before. Take your time and enjoy the experience. You'll find that the bees are docile and cooperative. Read the instructions several times until you become familiar and comfortable with the steps. Do a dry run before your girls arrive. The photo illustrations provide a helpful visual cue — after all, a picture is worth a thousand words!

TIP

Be sure to check out the video on putting bees in your hive at www.dummies.com/go/beekeepingfd4e.

When I hived my first package of bees, I had my wife standing by with the instructions, reading them to me one step at a time. What teamwork!

Ideally, hive your bees in the late afternoon on the day you pick them up or the next afternoon. Pick a clear, mild day with little or no wind. If it's raining and cold, wait a day. If you absolutely must, you can wait two to three days to put them in the hive, but make certain you spray them two or three times a day with sugar syrup while they're waiting to be introduced to their new home. Don't wait more than a few days to hive them. The sooner it's done, the better. Chances are they've already been cooped up in that box for several days before arriving in your yard.

Whenever I hive a package of bees, I always invite friends and neighbors to witness the adventure. They provide great moral support, and it gives them a chance to see firsthand how gentle the bees actually are. Ask someone to bring a camera. You'll love having the photos for your scrapbook!

## Hiving steps for Langstroth hives and Steps 1–7 for Top Bar hives

To hive your bees, follow these steps in the order they are given. Generally speaking, the first part of the process is the same whether you are using Langstroth hives or Top Bar hives. If you're using Top Bar hives, follow the first seven steps and then skip to the next section to complete the process.

**1.** **Thirty minutes before hiving, spray your bees rather heavily with sugar syrup.**

But don't drown them with syrup. Use common sense, and they'll be fine. The syrup not only makes for a tasty treat, the sticky stuff makes it harder for the bees to fly. They'll be good as new after they lick the syrup off each other.

*Courtesy of Howland Blackiston*

**2.** **Using your hive tool, pry the wood cover off the package.**

Pull the nails or staples out of the cover and keep the wood cover handy. You will need it again later.

*Courtesy of Howland Blackiston*

3. **Jar the package down sharply on its bottom so your bees fall to the bottom of the package.**

Don't worry. It doesn't hurt them!

*Courtesy of Howland Blackiston*

4. **Remove the feeding can of syrup from the package and the queen cage.**

Loosely replace the wood cover (without the staples) to keep the bees from flying out.

*Courtesy of Howland Blackiston*

5. **Examine the queen cage. See the queen?**

She's in there with a few attendants. Is she okay? In rare cases, she may have died in transit. If that's the case, go ahead with the installation as if everything were okay. But call your supplier to order a replacement queen (there should be no charge). Your colony will be fine while you wait for your replacement queen.

*Courtesy of Howland Blackiston*

6. **Remove the cork at one end of the cage so you can see the white candy in the hole.**

If the candy is missing, you can plug the hole with a small piece of marshmallow. Don't let the queen escape.

7. **If the queen cage lacks a metal strap, fashion a hanging bracket for the wooden queen cage out of two small frame nails bent at right angles.**

   Be sure to apply these nails at the candy end of the cage.

   *Note:* At this point, if you are installing your bees in a Top Bar hive, head to the section "Hiving Steps 8–14 for Top Bar hives."

8. **Prepare the hive by removing four or five of the frames, but keep them nearby.**

   Remember that at this point in time you're using only the lower deep hive body for your bees.

*Courtesy of Howland Blackiston*

Now hang the wooden queen cage (candy side up) between the frame that is closest to the open area and the frame next to it. The screen side of the cage can face toward the front or rear of the hive.

**TIP**

Some packages of bees come with instructions to hang the wooden queen cage with the candy side down. Don't do it. If one or more of the attendant bees in the cage dies, they will fall to the bottom and block the queen's escape. By having the candy side up, no workers will block the escape hole. If the queen came in a small plastic cage without attendants, the candy plug can go up or down.

9. **Spray your bees again, and jar the package down sharply so the bees drop to the bottom.**

   Toss away the cover — you're done with it.

**10.** **Pour the bees into the hive.**

Pour (and shake) approximately half of the bees directly above and onto the hanging queen cage. Pour (and shake) the remaining bees into the open area created by the missing frames.

*Courtesy of Howland Blackiston*

**11.** **When the bees disperse a bit, gently replace all but one of the frames.**

Be careful not to crush the bees as you reinstall the frames. Be slow and gentle. That extra frame will be installed a week later when you remove the empty queen cage, allowing the space for this frame.

**12.** **Place the hive-top feeder on top of the hive.**

Fill the feeder with sugar syrup.

*Courtesy of Howland Blackiston*

**13.** Now place the outer cover on top of the hive-top feeder. You're almost done.

**14.** Insert your entrance reducer, leaving a one-finger opening for the bees to defend.

Leave the opening in this manner until the bees build up their numbers and can defend a larger hive entrance against intruders. This takes about four weeks. If an entrance reducer isn't used, use grass to close up all but an inch or two of the entrance.

*Courtesy of Howland Blackiston*

**TIP**

Place the entrance reducer so the openings face up. Doing so allows the bees to climb up over any dead bees that might otherwise clog the small entrance.

**15.** Leave your Langstroth hive alone for a week. Seriously. No peeking until this time next week.

Give the girls a chance to know and accept their new queen. If you open the hive too soon, they might actually kill the queen. Yikes!

## Hiving Steps 8–14 for Top Bar hives

Installing your bees in a Top Bar hive follows a similar procedure to the one outlined above. In fact, the initial preparations outlined in Steps 1 through 7 are identical. After completing Steps 1–7 in the preceding section, follow these steps:

**8.** Remove all the bars from the hive except for the first four that are nearest to the front entrance.

**TIP**

Do not start this without a full set of bars that will allow you to fill the entire Top Bar hive from front to back. Also, you will need a feeder of some kind and enough sugar syrup to fill the feeder (and keep it filled).

You will hang the queen cage in the space between the fourth and the remaining top bars.

9. **Attach enough wire to the queen cage to allow it to hang about an inch below the center starter strips on the underside of the top bars and still have enough wire to extend up and hook over the fourth bar.**

10. **Hang the queen on the fourth bar and pin the wire securely to the top of the bar (a push pin or staple does the trick).**

**REMEMBER**

If the queen came in a wooden cage with attendants, make sure the candy is up. If she came in a small plastic cage without attendants, the candy plug can go either up or down.

11. **Spray your bees again and jar the package down sharply so the bees drop to the bottom.**

Toss away the cover of the package — you're done with it.

12. **Pour the bees into the hive.**

Pour (and shake) approximately half of the bees directly above and onto the queen cage that is hanging on the fourth bar. Pour (and shake) the remaining bees into the open area created by the missing top bars.

13. **Replace all the bars so the hive is completely sealed and put on the top cover.**

At this point the bees will begin to gather around the queen cage and form a cluster. This can happen rapidly, but allow an hour before completing the next step.

14. **Remove the cover and enough bars so you can install a feeder.**

After the hour wait time, the bees should have gathered around the queen in a cluster. Position the feeder as close to that cluster as possible, leaving only an empty bar or two between the feeder and the cluster. The bees will begin to immediately draw comb and will work toward the feeder fast. Close the colony and allow the bees to work. Be sure to set the entrance opening to the smallest option.

## Watching your bees come and go from their new home

Congratulations! You're now officially a beekeeper. You've launched a wonderful new hobby that can give you a lifetime of enjoyment.

**TIP**

Use this first week to get to know your bees. Take a chair out to the hive and sit to the side of the entrance — about 2 to 3 feet from the hive (within reading distance). Watch the bees as they fly in and out of the hive. Some of the workers will return to the hive with pollen on their hind legs. Other bees will be fanning at the entrance, ventilating the hive or releasing a sweet pheromone into the air. This scent is unique to this hive and helps guide their foraging sisters back to their home. Can you spot the guard bees at the entrance? They're the ones alertly checking each bee as she returns to the hive. In the afternoon, do you see any drones? They are the male bees of the colony and are slightly larger and more barrel-shaped than the female worker bees. The loud, deep sound of their buzzing often distinguishes them from their sisters.

## KNOWING WHEN AND HOW TO USE THE ENTRANCE REDUCER

The entrance reducer of a Langstroth hive or the adjustable entry on a Top Bar hive is used for two primary reasons:

- To regulate the hive's temperature

- To restrict the opening so a new or weak colony can better defend the colony

That being the case, here are some guidelines:

- For a newly hived colony, leave the entrance reducer in place until approximately four weeks after you have hived your package of bees. Chances are after a couple of weeks you can position the entrance reducer so its next largest opening is used. After about eight weeks, you can remove the entrance reducer completely. By that time the colony should be strong enough to defend itself — and the weather should be warm enough to fully open the entrance.

- For an established colony, use the entrance reducer during long periods of cold weather (less than 40° Fahrenheit [4° Celsius]). It helps prevent heat from escaping from the hive. I prefer not to use the smallest opening because I find it too restrictive — bees that die from attrition can clog the small opening. As a general rule of thumb, remove the entrance reducer completely when daytime temperatures are above 60° Fahrenheit (15° Celsius).

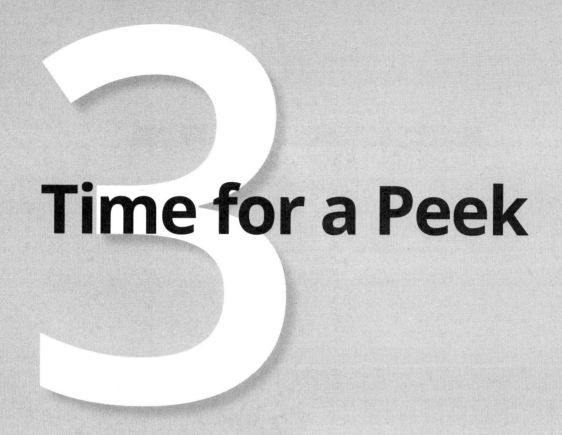

# 3

# Time for a Peek

Discover the tricks for successfully lighting your smoker and keeping it going for hours.

Read about the best and safest ways to inspect and enjoy your bees as well as how to maintain a healthy colony year-round.

Develop good habits right from the start by following tried-and-true "rules of the road."

Become familiar with basic inspection techniques, and appreciate what it is you are looking for — and why it is important.

Use the Beekeeper's Calendar to help you identify what activities you should be doing at a certain time of year, regardless of where you live in the world.

# Chapter **7**

# Opening Your Hive

T his is the moment you, as a new beekeeper, have been waiting for — that exhilarating experience when you take your first peek into the hive. You likely have a touch of fear, tempered by a sweeping wave of curiosity.

Put those fears aside. You'll soon discover visiting with your bees is an intoxicating experience that you eagerly look forward to. What you're about to see is simply fascinating. It's also one of the more tranquil and calming experiences that you can imagine: the warmth of the sun; the sweet smell of pollen, wax, and honey; the soothing hum of the hive. You're at one with nature. Your new friendship with your bees will reward you for many years to come.

**TIP**

The habits you develop in the beginning are likely to stick with you, so developing good habits early on is important. By getting familiar with the safe and proper way to inspect your hive and following suggested steps religiously in the beginning, you'll minimize any risks of injuring or antagonizing your bees. The techniques become second nature in no time. Down the road, you may find variations on the suggested methods that suit you better. Or helpful hints from other experienced beekeepers (there's more than one way to do beekeeping). That's okay. For now, just relax, move calmly, use good judgment, and enjoy the miracle of beekeeping.

# Establishing Visiting Hours

Ideally, open your hive on a nice, sunny day. Between 10 a.m. and 5 p.m. is best. Under those conditions, thousands of busy worker bees are out in the field. Avoid cold, windy, or rainy weather because that's when the entire colony is at home. With everyone in the hive, you'll probably find too many bees to deal with, particularly when you're just starting. In addition, the bees tend to be crankier when they can't get out of the house! You know how that is.

# Setting an Inspection Schedule

TIP

For the new beekeeper, once a week isn't too often to visit the bees. Use these frequent opportunities to find out more about the bees and their life cycles. Your first season is a time of discovery. You'll begin recognizing what's normal and what's not. You'll also become increasingly comfortable with manipulating the frames and working with the bees. So much so that inspecting your bees soon becomes second nature, and a quick peek at the entrance or under the lid is all that's needed to assure you that all is well. Beekeeping is as much an art as it is a science. Practice makes perfect.

Once you begin getting the hang of it, you needn't conduct more than six to eight inspections a year: Three or four visits in the early spring, one or two during the summer, and a couple of inspections at the end of the season are all that are absolutely necessary. But, hey, they're your bees, and I know you'll want to visit them regularly.

WARNING

Every day that you smoke the bees, open their hive, and pull it apart sets their productivity back a bit. It takes a day or two for life in the colony to return to normal. So if harvesting lots of honey is your objective, limit your inspections to once every few weeks.

This schedule doesn't apply to your first year when you need to gain greater experience by visiting the hive often.

As a beekeeper of a Top Bar hive, you are on a slightly different inspection schedule. Your immediate goal after installation of the bees is to get your bees started building straight comb. That will require some real early inspections, and possibly some comb management. Skip ahead to "Opening a Top Bar Hive" for a general overview of some important differences between opening a Langstroth hive versus a Top Bar hive.

# Preparing to Visit Your Langstroth or Top Bar Hive

The weekend has rolled around, and the weather's great (mild, sunny, and not much wind), so you've decided that you're going to pay the girls a visit. It's time to see what's going on in the hive. But you can't just dash out and tear the top off the hive. You have to get yourself ready for this special occasion. What will you wear? How will you approach them? What in the world do you do with all this new equipment?

In the upcoming section, I take you through the details of each step; they apply whether you're visiting a Langstroth or Top Bar hive.

**TIP**

You may want to read this chapter and the next one word for word. You may even want to read it a few times before having your first "close encounter." You may also want to take the book along on your first inspection, just in case you need some quick moral support. Better yet, coerce a friend or family member to go with you. That's what I did the first time. I had my wife reading from a book and prompting me through each step. At the time we didn't have an extra veil for her, so she hollered instructions at me from a safe distance.

## Making "non-scents" a part of personal hygiene

Forgive me for being personal, but you need to know that bees don't react well to bad body odor. So please don't inspect your bees when you're all sweaty after a morning jog. Take a shower first. Brush your teeth. On the other hand, don't try to smell too good, either!

**TIP**

Avoid using colognes, perfumes, or scented hairsprays. Sweet smells can attract more attention from the bees than you want.

**TIP**

Be sure to remove your leather watchband before visiting your bees. They don't like the smell of leather or wool, and these materials retain body odor. Removing your rings also is a good idea. It isn't that bees don't like pretty jewelry. But in the rare event that you take a sting on your hand, you don't want your fingers swell-ing up when you're wearing a decidedly nonexpandable ring.

## Getting dressed up and ready to go

Always wear your veil when you're inspecting your hive. Doing so keeps the bees away from your face and prevents them from getting tangled in your hair. For a discussion of the types of veils that are available, see Chapter 5.

**TIP**

If a bee ever gets under your veil, try not to panic. It isn't that big a deal. Her first instinct is to get out, so she's unlikely to sting you unless you squeeze her. To help her escape, simply walk away from the hive and slip off your veil. Don't remove your veil at the hive, and don't thrash around screaming and yelling. Doing so only upsets the bees, and the neighbors will think you've gone wacky.

New beekeepers need to wear a long-sleeved shirt. Light colors and smooth fabrics (like cotton) are best because bees don't like dark colors or the smell of wool or leather (material made from animals). If you aren't using commercially available beekeeping coveralls, use Velcro straps or rubber bands around the cuff of each pant leg and sleeve to keep your clothing bee-tight. Unless, of course, you think you might actually like having curious bees traveling up your sleeves and inside your trousers.

You can use gloves if you feel you absolutely must (see Chapter 5 for more information about gloves), but I encourage you not to develop that habit. Gloves are bulky. They impair your sense of touch and make your movements clumsy. When you're working with new colonies and early spring colonies, gloves aren't even necessary. These small, young, and gentle colonies are a delight to work with. Save your gloves for unfavorable weather, moving colonies around, or for use during the late summer and honey-harvesting time (when the colony's population is large and bees tend to be more defensive). But at all other times, I recommend that you leave the gloves at home or tucked into your pocket. Trust me. You'll thank me later.

**REMEMBER**

Colonies can be handled with far more dexterity and fewer injuries to bees (and you!) when you don't use gloves at all. Less injury to bees means a more docile colony.

# Lighting Your Smoker

The smoker is the beekeeper's best friend. Yet for many, keeping a smoker lit can be the trickiest part of beekeeping. It doesn't have to be. What you're trying to achieve is enough thick, cool smoke to last throughout your inspection. You certainly don't want your smoker to poop out as soon as you've opened the hive.

Begin with a loosely crumpled piece of newspaper about the size of a tennis ball. Light the paper and place it in the bottom of the smoker. Nest it in place using your hive tool. Gently squeeze the bellows a few times until you're sure that the paper is burning with a flame.

Add dry matchstick-size kindling, pumping the bellows as you do. As it ignites (you'll hear it crackling), slowly add increasingly thicker kindling. Ultimately, the fattest of your twigs will be about as thick as your thumb. None of the kindling need be more than 4 or 5 inches in length. The kindling needs to fill three-quarters of the smoker and must be thoroughly packed from side to side. Using your hive tool, occasionally stoke the fire. Keep pumping. When your kindling has been burning for about 10 minutes, and embers are glowing, it's time to add the real fuel.

Use a fuel that burns slowly and gives off lots of smoke. I'm partial to dry wood chips, a wad of burlap, or a fistful of hemp baling twine. But dry leaves and dry pine needles do nicely. You can also purchase *smoker fuel* (usually cartridges of compacted raw cotton fibers or nuggets of wood) from beekeeping supply stores. They work well, too. The bees really don't care what you use — but avoid using anything synthetic or potentially toxic. Figure 7-1 shows a smoker and readily available kindling and fuel.

**FIGURE 7-1:**
A smoker with all the "ingredients" is ready to load with paper, kindling of various sizes, and some natural fiber baling twine.

*Courtesy of Howland Blackiston*

You can learn how to light a smoker in the video at `www.dummies.com/go/beekeepingfd4e`. Check it out and see how I get my smoker to burn for hours!

The compressed wood pellets used as fuel in pellet stoves make fairly decent smoker fuel. Start with crumpled newspaper and a few twigs, then add a fistful or two of the wood pellets. Your smoker will produce thick, white smoke for hours!

**TIP**

Keep a box of kindling and fuel with your other beekeeping equipment. Having this readily available saves time on the days you plan to visit the hive.

Pack the smoker right to the top with your preferred fuel as you continue to gently pump the bellows. Gently tap it all down using your hive tool. When billows of thick, cool white smoke emerge, close the top. Pump the bellows a few more times. Use a long, slow pumping method when working the bellows, rather than short, quick puffs. Doing so produces more and thicker smoke than short puffs (see Figure 7-2).

**FIGURE 7-2:**
A smoking smoker — the beekeeper's best friend.

*Courtesy of Howland Blackiston*

## KEEPING YOUR SMOKER CLEAN

A good question that I'm frequently asked is: "My smoker is all gummed up and needs a good cleaning. How do I clean it?"

After a season or two, the inside of a smoker can become thickly coated with black, gummy tar. I've found the best way to clean it is by burning the tar out of it — literally. Like a self-cleaning oven, you need a great amount of heat. I've had success using a small propane blowtorch. You can purchase one at any hardware store. Just apply the flame to the black tar coating inside of your smoker. Keep blasting away. Soon the tar ignites, glows a fiery orange, and then turns to a powdery ash. Turn off the blowtorch. Once the metal smoker cools, you can easily knock the ash out of the smoker. Clean as a whistle!

Congratulations! You're now ready to approach the hive. Your smoker should remain lit for an hour or more.

**WARNING**

Make certain the smoke coming out of the smoker is "cool." You don't want to approach the hive with a smoker that is producing a blast furnace of smoke, fire, and sparks. Place your hand in front of the chimney as you gently work the bellows and feel the temperature of the smoke. If it feels comfortable to you, it will to the bees, too.

# Opening a Langstroth Hive

You're all suited up, and you have your smoker and hive tool. Perfect! Be sure to bring along an old towel (I'll explain why later in the "Removing the hive-top feeder" section). So now the moment of truth has arrived.

**TIP**

To see a video of me demonstrating how to open a hive, go to www.dummies.com/go/beekeepingfd4e. The steps are the same as the ones outlined here.

Approach your Langstroth hive from the side or rear. Avoid walking right in front of it because the bees shooting out the entrance will collide with you. As you approach the hive, take a moment to observe the bees and then ask yourself, "In what direction are they leaving the hive?" Usually it's straight ahead, but if they're darting to the left or right, approach the hive from the opposite side.

Follow these steps to open a Langstroth hive:

**1.** **Standing at the side and with your smoker 2 or 3 feet from the entrance, blow several puffs of thick, cool smoke into the hive's entrance (see Figure 7-3).**

Four good puffs of smoke should do fine. Use good judgment. Don't over-smoke them. You're not trying to asphyxiate the bees; you simply want to let the guard bees know you're there.

**2.** **Still standing at the side of the hive? Good. Now lift one long edge of the outer cover an inch or so, and blow a few puffs of smoke into the hive (see Figure 7-4).**

Ease the top back down and wait 30 seconds or so. Doing so gives the smoke time to work its way down into the hive. These puffs are for the benefit of any guard bees at the top of the hive.

# WHAT DOES THE SMOKE DO?

Smoke changes the bees' behavior and prevents them from turning defensive during inspections. You may ask, "Why?" One explanation I was told years ago is that it tricks bees into thinking there's a fire. In nature bees make their homes in hollow trees. So a forest fire would be a devastating event. Smelling the smoke, the bees fan furiously to keep the hive cool. They also begin collecting their most precious commodity — honey, engorging their *honey stomachs* with it in the event they must abandon ship and move to a new and safer home. With all the commotion, they become quite oblivious to the beekeeper. And when the inspection is complete and the crisis passes, the bees return the honey to the comb. That way, nothing is lost.

But I think another explanation is more likely. The smoke masks the alarm pheromones given off by worker bees when the hive is opened. Ordinarily, these alarm pheromones trigger defensive action on the part of the colony, but the smoke confounds the bees' ability to communicate danger.

In any event, smoking the bees works. Don't even think about opening a hive without first smoking it. It's a tempting shortcut that may work when your colony is brand-new, small, and young. But after that, it's a shortcut you'll try only once.

**FIGURE 7-3:**
Approach the hive from the side and blow a few puffs of smoke into the entrance to calm the guard bees.

*Courtesy of Howland Blackiston*

*Courtesy of Howland Blackiston*

3. **Put your smoker down and, using both hands, slowly remove the outer cover.**

   Lift it straight up and off the hive. Set the cover upside down on the ground (with the flat metal top resting on the ground and its underside facing skyward).

Your next step depends on whether you're still feeding your bees at the time of the inspection. If no hive-top feeder is on the colony, skip ahead to the section "Removing the inner cover."

Check out my video at www.dummies.com/go/beekeepingfd4e to see how I go about smoking my bees and opening my hive for inspection.

TIP

## Removing the hive-top feeder

If you're using a hive-top feeder, you'll need to remove it before inspecting your hive. To do so, follow these steps:

1. **Using your smoker, puff some smoke through the screened access and down into the hive (see Figure 7-5).**

*Courtesy of Howland Blackiston*

2. **Hive parts often stick together, so use the flat end of your hive tool to gently pry the feeder from the hive body (see Figure 7-6).**

   Do this slowly, being careful not to pop the parts apart with a loud "snap." That only alarms the bees.

TIP

   Here's a useful trick. Use one hand to gently press down on the feeder, while prying the feeder loose with the hive tool in your other hand. This *counter-balance* of effort minimizes the possibility of the two parts suddenly popping apart with a loud "snap."

3. **Loosen one side of the feeder and then walk around the back — not the front — and loosen the other side.**

4. **Blow a few puffs of smoke into the crack created by your hive tool as you pry loose the feeder.**

5. **Wait 30 seconds and completely remove the hive-top feeder.**

   Be careful not to spill any syrup. Set the feeder down on the outer cover that is on the ground.

TIP

   Positioning the feeder at a right angle to the cover when you set it down results in only two points of contact and makes it less likely that you'll crush any bees that remain on the underside of the feeder. Always be gentle with them, and they'll always be gentle with you!

**FIGURE 7-6:**
Use your hive tool as a lever to ease apart hive parts.

*Courtesy of Howland Blackiston*

**WARNING**

Remember the old towel I talk about earlier in this chapter? This is where it comes in. If syrup remains in the feeder, completely cover it with the towel (alternatively, you can use a small plank of plywood or a scrap of carpeting). Syrup left in the open attracts the bees — big time! You don't want to set off *robbing.* That's a nasty situation where bees go into a wild frenzy after finding free sweets (see Chapter 10). Open containers of syrup (or honey for that matter) also can attract bees from other colonies. All the gorging bees wind up whipped into such a lather that they begin robbing honey from your hive. War breaks out, and hundreds or even thousands of bees can be killed by the robbing tribe. Enough said! A good rule of thumb: *Never* leave syrup in the wide open. Keep it out of reach!

## Removing the inner cover

If you're *not* using a hive-top feeder, you'll need to remove the inner cover (an inner cover is always used *unless* a top feeder is on the hive). Removing the inner cover is much like removing the top feeder. Follow these steps:

**1.** Puff smoke through the oval hole and down into the hive.

**2.** Using the flat end of your hive tool, gently release the inner cover from the hive body (see Figure 7-7).

Loosen one side and then walk behind the hive and loosen the other side. Pry slowly, being careful not to pop the parts apart with a loud "crack."

**FIGURE 7-7:** Direct some smoke through the hole in the inner cover — wait half a minute and then use your hive tool to release and remove the inner cover from the hive body.

*Courtesy of Howland Blackiston*

3. **Blow a few puffs of smoke into the crack created by your hive tool as you pry up the inner cover.**

4. **Wait 30 seconds and then completely remove the inner cover.**

   Set it down on the outer cover that's now on the ground, or simply lean it up against a corner of your hive. Careful! Don't crush any bees that may still be on the inner cover.

# Opening a Top Bar Hive

With a Top Bar hive, there is no difference regarding how you prepare for a visit and smoke your bees. So here I deal with those details that *are* a little different when dealing with a Top Bar hive.

**WARNING**

The comb in a Top Bar hive is very fragile. Be very, very careful when handling it. Never tilt the comb forward or backward, allowing the comb surface to be positioned horizontally, against the force of gravity. Figure 7-8 shows the three safe positions for handling Top Bar hive comb. Because there are no bottom or side supports, the comb in a Top Bar hive requires constant support. This is particularly the case with new comb — but the warning applies to all Tob Bar hive comb.

**FIGURE 7-8:** These are the three positions you must maintain when handling Top Bar hive comb. Any other position will result in the comb detaching from the bar.

Top Bar hives are opened from the rear of the hive, meaning the end farthest from the entrance. That's because the bars farthest from the entrance are the last ones on which the bees draw wax comb. This approach is less disturbing to the colony because the bars will be relatively free of comb and bees until the colony really gets established. Also, because the older bees and the guard bees are up front near the entrance, you give yourself a little buffer from disturbing these more defensive bees at the start of the inspection process.

If the entrance is in the middle, you have two ends to choose from when approaching and opening the hive. Either one will do.

Follow these steps to open a Top Bar hive:

**1. Smoke the entrance just as with any hive.**

   Pause a moment or two before proceeding to let the smoke do its magic.

2. **Remove the top cover.**

   Careful — it's heavy. Place the cover nearby on the ground. With a Top Bar hive, there is no "inner cover" to remove, so the top bars are exposed at this point.

**WARNING**

   Once you remove the cover, the long flat surface of top bars seems like a natural place to rest your smoker — but don't do it. The bottom of your smoker is hot enough to damage or melt the wax hanging just below on the bars. Put the smoker down on the ground when you are not using it.

3. **Remove the back bars until the feeder is exposed and then remove the feeder, being careful not to spill any syrup.**

**TIP**

   Cover the open feeder to avoid the possibility that other bees in your area are attracted to it.

# The Hive's Open! Now What?

Whew! The hive is officially open. Relax and take a deep breath. You should see lots of beautiful bees! Here's what to do next:

1. **Time for the smoker again.**

   From 1 or 2 feet away, and standing at the rear or side of the hive, blow several puffs of cool smoke between the frames and down into the hives. Pumping the bellows in long, slow puffs, rather than short, quick ones, make sure that the breeze isn't preventing smoke from going into the spaces between the frames. Watch the bees. Most of them will retreat down into the hive.

2. **Now you can begin your inspection (see Chapter 8).**

**WARNING**

   Although you have much to do, you don't want to keep the hive open for more than 15 to 20 minutes (even less if the weather is cooler than 55 degrees Fahrenheit [13 degrees Celsius]). But don't rush at the expense of being careful! Clumsiness results in injury to bees, and that can lead to stings. Be gentle with the ladies!

In Chapter 8, I explain exactly what you should look for when the hive's wide open.

# Chapter **8**

# What to Expect When You're Inspecting

Peering out through your veil with your cuffs strapped shut and your smoker lit, you've opened your hive and see that it's bustling with bees. But what exactly are you looking for?

Understanding when to look and what to look for makes the difference between being a "beekeeper" and a "bee*haver.*" Anyone can *have* a hive of bees, but your goal as a bee*keeper* is to help these little creatures along. Understand their needs. Try to anticipate problems. Give them the room they need before they actually need it. See that they have comb in which to store honey before the nectar starts to flow. Feed them when food supplies are low. Nurse them back to health when illness strikes. Get them ready for winter before the weather turns cold. In return, your bees will reward you with many years of enjoyment and copious crops of sweet, golden honey.

You always follow certain procedures, and you always look for certain things when inspecting your hive. After a few visits, the mechanics become second nature, and you can concentrate on enjoying the miraculous discoveries that await you. In this chapter, I give you some pointers that make each inspection easy. And you look at how the procedures differ slightly when inspecting Langstroth versus Top Bar hives.

# Keeping a Journal

There are a couple of good reasons to keep a journal of your hive inspections, regardless of whether you're inspecting a Langstroth hive, Top Bar hive, or any other kind of hive. First, the journal can serve as a helpful checklist to ensure you perform a complete inspection. Second, the journal provides you with a historical record of what you noted on the previous visits and thus serves as a useful way to compare the colony's progress from one inspection to the next. You can also add reminders in the journal: "On next visit replace the broken entrance reducer," or "This hive will need a honey super added in another week." There's a helpful beekeeper's checklist in the Appendix section of this book.

## KEEPING NOTES GOES VIRAL

Having a written record of your inspections is very helpful. In fact, it's imperative if you have multiple colonies to keep track of. And while pen and paper work great, in these days of smartphones, tablets, and interactive databases, there's an even niftier option out there. Two computer scientists (who also happen to be beekeepers) came up with an easy-to-use online application called Hive Tracks. The app prompts what needs to be done and when, and then securely stores inspection notes, harvest results, observations, photos, queen performance, colony health, and many other details. The 30-day trial version is free, and there is a relatively modest annual fee for the software. See the special discount coupon offer from Hive Tracks at the back of this book. And by all means, have a look at their website at https://hivetracks.com.

*Image courtesy of Hive Tracks*

# Inspecting a Langstroth Hive

Once you have smoked and opened your hive (see Chapter 7), it's time for a peek inside. So take a deep breath, relax, and follow these sequential steps. You're in for a treat as you enter your bees' secret world!

## Removing the first frame of your Langstroth hive

Always begin your inspection of the Langstroth hive by removing the *first frame* or *wall frame.* That's the frame closest to the outer wall. Which wall? It doesn't matter. Pick a side of the hive to work from, and that determines your first frame. Here's how to proceed:

1. **Insert the curved end of your hive tool between the first and second frames, near one end of the frame's top bar (see Figure 8-1).**

**FIGURE 8-1:**
Use your hive tool to pry the wall frame loose before removing it.

*Courtesy of Howland Blackiston*

2. **Twist the tool to separate the frames from each other.**

   Your hand moves toward the center of the hive — not the end.

3. **Repeat this motion at the opposite end of the top bar.**

   The first frame should now be separated from the second frame.

4. **Using both hands, pick up the first frame by the end bars (see Figure 8-2).**

   Gently push any bees out of the way as you get a firm hold of the end bars. With the frame in both hands, *slowly* lift it straight up and out of the hive. Be careful not to roll or crush bees as you lift the frame. Easy does it!

FIGURE 8-2:
Carefully lift out the first frame and set it aside. Now you have room to manipulate the other frames.

*Courtesy of Howland Blackiston*

**WARNING**

You should never put your fingers on a frame without first noting where the bees are, because you don't want to crush any bees, and you don't want to get stung. Bees can be easily and safely coaxed away with a little smoke or by gently and slowly pushing them aside with your fingers.

Now that you've removed the first frame, gently rest it on the ground, leaning it vertically up against the hive on the side of the hive opposite from where you are standing. It's okay if bees are on it. They'll be fine. Or, if you have a frame rest (a handy accessory available at some beekeeping supply stores), use it to temporarily store the frame.

This is a basic and important first step every time you inspect a colony. The removal of this frame gives you a wide-open empty space in the hive for better manipulation of the remaining frames without squashing any bees.

## Working your way through the Langstroth hive

Using your hive tool, loosen frame two and move it into the open slot where frame one used to be. That gives you enough room to remove this second frame without the risk of injuring or crushing any bees. When you're done looking at this frame, return it to the hive, close to (but not touching) the wall. Alternatively, you can add it to the frame rest, if you are using one.

Work your way through all ten frames in this manner — moving the next frame to be inspected into the open slot. When you're done looking at a frame, always return it snugly against the frame previously inspected. Use your eyes to monitor progress as the frames are slowly and carefully nudged together.

As you gain experience, you'll find that you need not examine every single frame. You will become increasingly knowledgeable regarding what to look for and will be able to evaluate your hive by examining fewer frames.

**WARNING**

Be careful not to crush any bees when pushing the frames together. One of those bees may be the queen! Look between the frames to make sure the coast is clear before pushing the frames together. If bees are on the end bars and at risk of being crushed, use the flat end of your hive tool to coax them along. A single puff of smoke also urges them to move out of the way.

## Holding up frames for inspection

Holding and inspecting an individual frame the proper way is crucial. Be sure to stand with your back to the sun, with the light shining over your shoulder and onto the frame (see Figure 8-3). The sun illuminates details deep in the cells and helps you to better see eggs and small larvae. Here's an easy way to inspect both sides of the frame (Figure 8-4 illustrates the following steps):

**FIGURE 8-3:**
Hold frames firmly with the light source coming over your shoulder and onto the frame.

*Courtesy of Howland Blackiston*

1. **Hold the frame firmly by the tabs at either end of the top bar.**

   Get a good grip. The last thing you want to do is drop a frame covered with bees. Their retaliation for your clumsiness will be swift and, no doubt, memorable.

2. Turn the frame vertically.

3. Then turn the frame like a page of a book.

4. Now smoothly return it to the horizontal position, and you'll be viewing the opposite side of the frame.

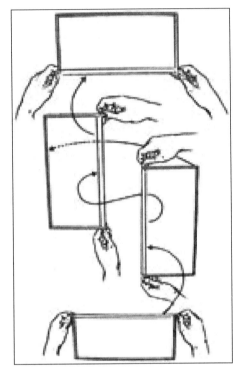

**FIGURE 8-4:**
The correct way to inspect both sides of a frame.

**TIP**

When inspecting frames, all your movements must be slow and deliberate. Change hand positions sparingly. Sliding your fingers across the frames as you reposition your hands is better than lifting your fingers and setting them down again because you may land on a bee. As you turn the frame, avoid any sudden and unnecessary centrifugal force that can disturb the bees or damage the comb.

## Knowing when it's time for more smoke

Regardless of what kind of hive you are inspecting, a few minutes into your inspection, you may notice that the bees all have lined up between the top bars like racehorses at the starting gate. Their little heads are all in a row between the frames. Kind of cute, aren't they? They're watching you. That's your signal to give the girls a few more puffs of smoke to disperse them again so you can continue with your inspection. See the "Understanding What to Always Look For" section later in the chapter to know what to inspect while the hive is open.

## Replacing Langstroth frames

After you've inspected your last frame, all but one of the frames should be in the hive and one leaning against it or hanging on the frame rest (the first frame you removed). Putting the first frame back in the hive means taking these steps:

1. **Slowly push the frames that are in the hive as a single unit toward the opposite wall of the hive.**

   That puts them back where they were when you started your inspection. Pushing them as a single unit keeps them snugly together and avoids crushing bees. Focus your eyes on the "point of contact" as you push the frames together. You're now left with the open slot from which the first frame was removed.

2. **Smoke the bees one last time to drive them down into the hive.**

3. **Pick up the frame that's outside the hive.**

   Are bees still on it? If so, with a sharp downward thrust, sharply knock one corner of the frame on the bottom board at the hive's entrance. The bees fall off the frame and begin walking into the entrance to the hive. With no bees remaining on your first frame, you can easily return it to the hive without the risk of crushing them.

4. **Ease the wall frame into the empty slot.**

   Slowly, please! Make certain that all the frames fit snugly together. Using your hive tool as a wedge, adjust the multi-frame unit so the space between the frames and the two outer walls is equal.

## Closing the Langstroth hive

You're almost finished. Follow these steps to close the hive:

1. **If you're using a hive-top feeder on a Langstroth hive, put it back in place immediately on top of the hive body.**

   Add more sugar syrup if the pantry is getting low. Now go to Step 4.

2. **If you're not using a hive-top feeder, replacing the inner cover comes next.**

   First remove any bees from the inner cover. Use a downward thrust and sharply knock one corner of the inner cover on the bottom board at the hive's entrance. Better yet, if there is a rock on the ground, use it as your hard surface rather than the bottom board.

3. **Place the inner cover back on the hive by sliding it in position from the rear of the hive so you don't crush any bees.**

   Very slowly slide it into place, and any bees along the top bars or on the edges of the hive will be pushed gently out of the way.

   Note that the notched ventilation hole is positioned upward and toward the front of the hive. This notched opening allows air to circulate and gives bees a top-floor entrance to the hive. Some manufacturers of bee equipment do not have this ventilation opening — I suggest getting one that does have this nice feature.

4. **Replace the outer cover (the final step).**

   Make sure the outer cover is free of any bees. Tap it sharply on the ground to free it of bees. From the rear of the hive, slide it along the inner cover, again, gently pushing any bees out of the way (use the bulldozer technique). Ease it into place, and adjust it so it sits firmly and level on the inner cover.

**REMEMBER**

Make sure the ventilation notch on the outer cover isn't blocked. From the rear of the hive, shove the outer cover toward the front of the hive. Doing so opens the notched ventilation hole in the inner cover and gives the bees airflow and an alternate entrance.

Congratulations! The bees once again are snugly in their home.

# Inspecting a Top Bar Hive

When inspecting your Top Bar hive, always remember that the comb is fragile and if handled without care can detach from the top bar. So be careful and your experience will be pleasant and rewarding.

Assuming you have just installed a package, your first inspections will be to ensure the following: The queen was released; the bees are drawing straight comb; and your feeder is full of syrup. New bees in a Top Bar hive will require continual feeding until your main nectar flow commences. Depending on where you live, you may be feeding your Top Bar colony for a while.

# Working your way through the Top Bar hive

**TIP**

If you have just established the Top Bar hive, you'll be feeding the colony for as long as it takes to get at least 10 to 15 bars drawn or until the colony stops taking syrup. So the first task will be to remove the feeder and then fill it after you are done.

After you've removed the feeder, you can see the top bars where the comb is being drawn. Starting at the most remote bar from the entrance, remove all the empty bars until you encounter the first bar being worked by the bees. Look for and remove any bridge comb connected to the sidewall.

Top Bar comb (and especially new comb) is used to store honey. It is almost always bridged near the top. *Bridging* is when bees attach a small amount of wax between the comb and the top of the sloped sidewall (see Figure 8-5). Before you attempt to lift a bar, this bridging wax will need to be unattached. It's not difficult. Just use your hive tool or a long thin knife and slice through it. You can now lift the bar to inspect it or slide it back out of the way.

**FIGURE 8-5:** Bridging. Notice the wax between the top of the comb and the sloped side.

*Courtesy of William Hesbach*

After you've removed the bridge, insert your hive tool between that end of that bar and the next to make a little space. Do the same on the other side, and the bar should be free to slide away or to lift up. After a bar is inspected, it can be slid into the empty space created by removing the blank bars so it's safely out of the way while the next bar is inspected.

**WARNING**

If you feel resistance as you begin to move the bar, stop and make certain that it's not attached in some other way to the next bar. If it is, you may have to cut that connection before the bar can be moved.

**TIP**

Mark one end of all your top bars to ensure that they are always replaced in the same orientation.

As you slide your bars to the back of the colony, keep the top bars touching so you end up with all the bars together. This may require that you move bees out of the way, because they are likely to be on the top-bar edges checking things out. This can be done in one of two ways. One way is to start by touching the very end of the bars together on one side and then slowly narrow the angled gap. This will encourage bees to leave the gap as the closing bar pressures them. The other way is to lift the bar slightly above the bar already in place and lower it by first making contact on a slight angle, moving it slowly down while the bees scurry away. Whatever you try, remember not to crush bees, because crushed bees will release an alarm pheromone that will change their temperament quickly . . . and for the worst.

**REMEMBER**

If you need to move a lot of bees out of the way at once, use your smoker. To know when it's appropriate to do so, check out the "Knowing when it's time for more smoke" section earlier in the chapter.

This technique will work well until a colony builds out all the bars and there is no empty space. When your colony reaches that point, you must put aside the first few bars you inspect to make the space you need. You can turn them upside down and rest them on their top bars out of the way on any flat surface, or fabricate a frame rest that will hold them safely out of the way. Remember to consider the safety of any rest area you choose and make certain it's not too hot, in direct sunlight for an extended period, or too cold.

**WARNING**

Never tap or bang the edge of top bar comb against anything to make the bees fly away, because you'll damage the comb. Use a bee brush to gently remove them instead.

If this is the first inspection after hiving the bees (when you are checking to see if your queen was released), move back the top bars until you arrive at the bar with the queen cage. Lift that bar and examine the cage. If the queen is released from the cage, remove it and replace the bar in the hive.

## Top Bar comb management

At this early stage in comb building, you need to begin comb management and ensure that the bees are building straight along the top-bar guides. Two days after installing the bees in your Top Bar hive, inspect the colony. Begin by observing what they are doing by removing just the bars over the feeder. If they haven't built enough comb for you to judge how straight they are building, close the colony and check it again the next day.

The actual management manipulation comes into play if the bees are not building straight along the guide. In that case, you need to lift out the bar or bars and straighten the comb by gently repositioning the comb in alignment with the center guide.

If your bees are the type that will attach comb off the guide, it's usually evident right away (see Figure 8-6). This misalignment is known as *hooking*. Caught early, hooking can be corrected and the bees will continue to build straight. If you allow hooking combs to continue, they will eventually cross the adjacent bar, which will make future inspections impossible without great difficulty. A little comb management at this point will make your Top Bar hive experience much more enjoyable.

**TIP**

Bars that use a strip of foundation as a guide seem to help keep the bees building straight.

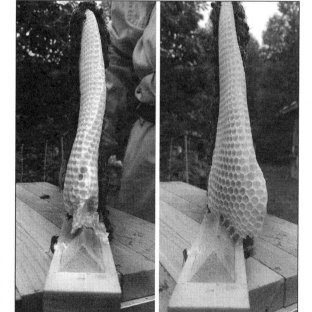

**FIGURE 8-6:**
Notice how this comb is hooked in the first figure and straightened in the next. It's a little messy but has a big payoff.

*Courtesy of William Hesbach*

## Looking into Top Bar cells

The cells on a Top Bar hive comb are sometimes, but not always, built with a slight upward angle. When they are, you must look down into the cell bottoms, keeping in mind that you can't tilt the frame, but rather must lift the frame past your gaze with the light source behind you. In some cases it's easier to see down to the bottom of cells when the frame is flipped upside-down and held above your head.

See the "Understanding What to Always Look For" section later in the chapter to know what else to inspect while the hive is open.

## Replacing the top bars and closing the hive

With the inspection complete, it's now time to start reassembling the hive. If you have been sliding bars into the empty space, you can now begin to slide them forward in groups, leaving the bars together as you go. Continue until all the inspected bars are back in their original space. Once all the bars are in place, you can fill the feeder, put in all the empty bars, and replace the cover. You're done! Time for a tea and honey break.

# Understanding What to Always Look For

Each time you visit your Langstroth or Top Bar hive, be aware of the things that you always must look for. Virtually all inspections are to determine the health and productivity of the colony. The specifics of what you're looking for vary somewhat, depending on the time of year. But some universal rules of the road apply to every hive visit.

## Checking for your queen

Every time you visit your hive, you're looking for indications that the queen is alive and well and laying eggs. If you actually see her, that's great and reassuring! But finding the queen becomes increasingly difficult as the colony becomes larger and more crowded. So how can you tell whether she's there?

Rather than spending all that time trying to see the queen, look for *eggs.* Although they're tiny, finding the eggs is much easier than locating a single queen in a hive containing tens of thousands of bees. Look for eggs on a bright, sunny day. Hold the comb at a slight angle and with the sun shining over your shoulder. This illuminates the deep recesses of the cells. The eggs are translucent white, resembling minuscule grains of rice.

**WARNING**

If you see more than a single egg in a cell, you may have a problem — drone-laying worker bees. This conundrum and its solution are covered in Chapter 10.

Binocular magnifiers (used by hobbyists and watchmakers) are better than reading glasses. You can see an image of this gadget in Chapter 2. These magnifiers can be worn under your veil and tipped out of the way when you aren't using them. They make egg spotting easier and give you a whole new perspective on the tiny wonders in the colony. Alternatively, you can make use of a conventional magnifying glass.

## Storing food; raising brood

Each deep Langstroth frame of comb contains about 7,000 cells (3,500 on each side). However, if you have a Top Bar hive, the number of cells will vary. That's because Top Bar comb is trapezoidal and has fewer cells as the comb tapers down. The number of cells varies depending on the width and depth of the Top Bar box.

Honey bees use their hexagonal cells for storing food and raising brood. When you inspect your colony, noting what's going on in those cells is important because it helps you judge the performance and health of your bees. Ask yourself: Is there ample pollen and nectar? Are there lots of eggs and brood? Does the condition of the wax cappings over the brood look *normal* — or are the cappings perforated and sunken in (see Chapters 10 and 12 for tips on recognizing unhealthy situations)?

## Inspecting the brood pattern

Examining the brood pattern is an important part of your inspections. A tight, compact brood pattern with the majority of cells of the same age in an area is indicative of a good, healthy queen (see this book's color insert). Conversely, a spotty brood pattern (many empty cells with scattered, mixed-age cells of eggs, larvae, or capped brood) is an indication that you have an old or sick queen and may need to replace her. How does the capped brood look? These are cells that the bees have capped with a tan wax. The tan cappings are porous and enable the developing larvae within to breathe. The cappings should be smooth and slightly convex. Sunken-in (concave) or perforated cappings indicate a problem. See Chapter 12 for more information about how to recognize the telltale signs of brood disease.

*Capped brood* refers to larvae cells that have been capped with a wax cover, enabling the larvae to spin cocoons within and turn into pupae.

## Recognizing foodstuffs

Learn to identify the different materials collected by your bees and stored in the cells. They'll pack pollen in some of the cells. Pollen comes in many different colors: orange, yellow, brown, gray, blue, and so on. The color insert section of this book has a photo showing a wide array of pollen colors stored in the cells of a comb.

As you inspect the cells further, you'll see non-capped cells with something "wet" in them. It may be nectar. Or it may be water. Bees use large amounts of water to cool the hive during hot weather.

# Your New Colony's First Eight Weeks

For the newly hived colony, some specific beekeeping tasks are unique to the first few weeks of your first season. When you do any inspection, the general method for smoking, opening, and removing the frames is the same as the methods given in the "Inspecting a Langstroth Hive" and "Inspecting a Top Bar Hive" sections earlier in this chapter.

## Checking in: A week after hiving your bees

After putting your package of bees in the hive, you'll be impatient to look inside to see what's happening. Resist the temptation! You must wait one full week before opening the hive. The colony needs this first uninterrupted week for accepting its new queen. Any premature disturbance to the hive can result in the colony rejecting her. The colony may even kill her, thinking the disturbance is somehow her fault. Play it safe and leave the hive alone for one week. During that time, worker bees eat through the candy and release the queen from her cage. She becomes the accepted leader of the colony.

**REMEMBER**

As mentioned in Chapter 7, conduct your first inspection on a mild, sunny day (55 degrees Fahrenheit [13 degrees Celsius] or more) with little or no wind. As always, visit your hive sometime between 10 a.m. and 5 p.m.

Smoke and open your hive, and remove the first frame. Other than a few occasional bees, not much will be happening on this frame. In all likelihood, the bees haven't had time to draw the foundation into honeycomb.

As you continue your inspection of each subsequent frame, you should begin to see more and more going on. Toward the center of the hive you should see that the girls have been busily drawing out the wax foundation into honeycomb.

## Verifying that the queen was released

When you reach the two frames sandwiching the queen cage, look down in the hole where the candy plug was. If the candy is gone, that's wonderful! It means worker bees have chewed through the candy plug and released the queen. Remove the cage and peek inside. Confirm that the queen has been released. Place the cage near the entrance so any worker bees exploring in the cage find their way back into the hive.

## Removing any burr comb

You're likely to find that industrious bees have built lots of *burr comb* (sometimes called *natural comb, wild comb,* or *brace comb*) in the gap created by the queen cage. You may find comb on and around the queen cage itself. Although it's a beautiful bit of engineering, you must remove this bright white comb of perfectly symmetrical cells. Failing to do so is sure to create all kinds of headaches for you later in the season.

TIP

In all likelihood this wild comb will contain eggs (it's the first place the queen lays eggs). Spotting those eggs will confirm you have a laying queen in the hive.

Use your hive tool to sever the burr comb where it's connected to the frames and slowly lift the comb straight up and out of the hive. You'll probably find it covered with worker bees.

Examine the comb to ensure the queen isn't on it. If she is, you must gently remove her from the comb and place her back into the hive. Queens are quite easy to handle, and although they have a stinger, they're not inclined to use it. Simply wet your thumb and index finger and gently grasp her by her wings. She moves quickly, so it may take a few tries. Don't jab at her, but rather treat her as if she were made of eggshells. Easy does it!

Removing the bees from this burr comb is a good idea for new beekeepers — one of those bees just might be the queen. If you are inspecting a Langstroth frame, one or two good shakes dislodge the bees from the burr comb. Shaking bees loose is a technique that will come in handy many times in the future. *Shaking* is a sharp downward motion with an abrupt halt just above the hive. See an illustration of this technique in Chapter 17. But this is not a technique I advise for a Top Bar hive. Top Bar comb is too delicate to handle the jolt. For Top Bar hives, use a bee brush to coax bees from the comb.

Save the natural comb to study at your leisure back at home. Look for eggs because the queen often starts laying on this comb. It makes a great show-and-tell for children! And you can always use beeswax to make cool things, like candles, furniture polish, and cosmetics recipes.

### Looking for eggs

Taking a close look at the frames that were near the queen cage, what do you see? Pollen? Nectar? Great! Do you see any eggs? They're the primary things you're looking for during your first inspection (see "Understanding What to Always Look For" earlier in this chapter). When eggs are present, you know the queen already is at work. That's all you need to find out on this first inspection. Close things up and leave the bees alone for another week. Be satisfied that all is well. Because the weather likely is still cool, you don't want to expose the new colony to the elements for too long.

WARNING

If no queen or eggs can be found, you may have a problem. In this abnormal situation, wait another few days and check once again. Seeing eggs is evidence enough that you have a queen, but if you still find no evidence of the queen, you need to order a new one from your bee supplier. The colony will do okay while awaiting its new queen, which you'll introduce exactly as you did the original: by hanging the cage between two frames (see Chapter 6) and leaving the bees alone for a full week.

TIP

New beekeepers often have a really hard time finding eggs. Before you give up hope as to whether there are any, look again. If you still don't see them, look again. And use a magnifier. Chances are they're there, and you just haven't gained experience as to what you're looking at. Remember, they are very, very small.

### Replacing the missing frame of the Langstroth

The missing frame is the one that you removed when you originally hived your package. It now becomes your *wall frame.*

### Providing more syrup

When necessary, replenish your feeder with more sugar syrup. The recipe for sugar syrup can be found in Chapter 6.

## The second and third weeks

On that first visit, you were looking for evidence that the queen had been released and was laying eggs. During the inspections that you conduct two and three weeks after hiving your package, you're trying to determine how well the queen is performing. By now there are a lot of new things to see and admire.

Following standard procedure, smoke, open the hive, and remove frames or bars one by one for inspection. Work your way toward the center of the hive. As always, look for eggs. They're your ongoing assurance that the queen is in residence.

Note that the bees have drawn more honeycomb. In a Langstroth hive, they work from the center outward, so the outer five to six frames haven't likely been drawn out yet. That's normal.

## Looking for larvae

By the second week you can easily see larvae in various stages of development (see Figure 8-7). They should be bright white and glistening like snowy white shrimp! Looking closely, you may even witness a larva moving in its cell or spot a worker bee feeding one.

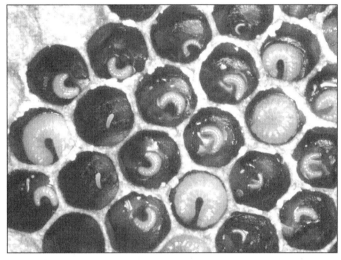

**FIGURE 8-7:** Larvae go through various stages of development. Also, note eggs in this photo.

*Courtesy of Howland Blackiston*

## Evaluating your queen

Estimate how many eggs her majesty is laying. One good way to tell is if you have one or two frames with both sides ¾-filled with eggs and larvae. That means your queen is doing a super-fantastic job. Congratulations!

If you have one or two combs with only one side filled, she's doing moderately well. If you find fewer than that, she's doing poorly, and you need to consider replacing her as soon as possible. See Chapter 10 for instructions on how to replace your queen.

## Hunting for capped brood

By the third week you'll begin seeing capped brood — the final stage of the bees' metamorphosis. *Capped brood* are light tan in color, but note that the brood cappings on older comb are a darker tan or even dark brown. The capped brood are located on comb that is closest to the center of the hive. Cells with eggs and larvae are on the adjacent comb.

TIP

An excellent queen lays eggs in nearly every cell, skipping few cells along the way and resulting in a pattern of eggs, larvae, and capped brood that is tightly packed together, stretching all the way across most of the comb.

You'll also notice a crescent of pollen above each capped brood and a crescent of nectar or capped honey above the pollen. This is a picture-perfect situation.

WARNING

A spotty and loose brood pattern also can be evidence of a problem. You may have a poor queen, in which case she should be replaced as soon as possible. Sunken or perforated brood cappings may be evidence of brood disease, in which case you must diagnose the cause and take steps to resolve it. (See Chapter 12 for more about bee diseases and remedies.)

## Looking for supersedure cells

The third week also is when you need to start looking for *supersedure* cells (also called queen cells). *Supersedure* is a natural occurrence when a colony replaces an old or ailing queen with a new queen. The bees create supersedure cells if they believe their queen is not performing up to par. These peanut-shaped appendages are an indication that the colony may be planning to replace (or supersede) the queen. Queen cells, often only one or two in number and located on the upper two-thirds of the comb, are most likely supersedure cells (see Figure 8-8). On the other hand, queen cells located on the lower third of the comb are likely not supersedure cells, especially if numerous, but are called swarm cells, which are discussed later in this chapter. *Note:* Swarming seldom is a problem with a new hive this early in the season.

The bees create *swarm cells* to raise a new queen in preparation for the act of *swarming.* This usually happens when conditions in the hive become too crowded. The colony decides to split in half — with half the population leaving the hive (swarming) with the old queen and the remaining half staying behind with the makings for a new queen (the ones that are developing in the queen cells, or swarm cells).

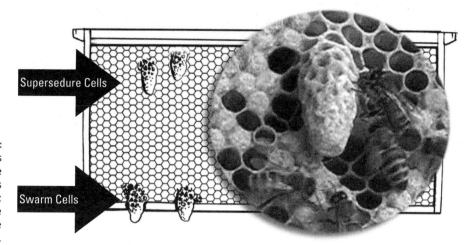

Courtesy of USDA-ARS, Stephen Ausmus

**FIGURE 8-8:**
Supersedure cells are located in the upper two-thirds of the comb; swarm cells are located along the bottom third.

**WARNING**

If you spot supersedure cells, especially early in the development of your new colony, you need to order a new queen, because giving the bees a new queen is usually better than letting them create one on their own. Why? Because you'll lose less time and guarantee a desirable lineage.

### Provide more syrup

Check every week to see whether your hive's feeder has enough sugar syrup. Replenish as needed. The recipe for sugar syrup is in Chapter 6. Just pour it into the feeder.

## Weeks four through eight

Things really are buzzing (or at least they should be) now that a month has passed since you hived your bees.

Perform your inspection as always, looking for evidence of the queen (eggs) and a good pattern of capped brood, pollen, and capped honey.

If you are using a Top Bar hive, the size of the hive remains constant as the colony grows. But if you are using a Langstroth hive, you will want to add additional space as the colony grows.

## Adding a second deep-hive body to your Langstroth hive

If all's well, by the end of the fourth week the bees have drawn nearly all the foundation into comb. They've added wax produced in their wax glands to the foundation, creating the hexagonal comb cells in which they store pollen, honey, and brood. When seven of the ten frames are drawn into the comb, you want to add your second deep-hive body (see Chapter 4 for more information about hive parts). Anticipate the need for this addition because timing is important. If you wait too long, the colony may grow too fast (with up to 1,000 new bees emerging every day!), become overcrowded, and eventually swarm. Add the second deep hive body too early and the colony below loses heat, and the brood may become chilled and die.

When adding the second deep-hive body becomes necessary, follow these steps:

1. **Smoke your hive as usual.**

2. **Remove the outer cover and the hive-top feeder (or the inner cover, if one is being used).**

3. **Place the second deep directly on top of the original hive body.**

4. **Fill the new second story with ten frames and foundation (see Chapter 4).**

5. **Put the hive-top feeder directly on top of the new upper deep and below the outer cover.**

   Replenish sugar syrup if needed.

6. **Replace the outer cover.**

The upper deep will be used during the early summer for raising brood. But later on it serves as the food chamber for storing honey and pollen for the upcoming winter season.

## Witnessing a miracle!

By the fifth week the combs are jampacked with eggs, larvae, capped brood, pollen, and honey. Look carefully at the capped brood. You may see a miracle in the making. Watch for movement under the capping. A new bee is about to emerge! She'll chew her way out of the cell and crawl out (see Figure 8-9). At first, she totters about, yet she quickly learns how to use her legs. She appears lighter in color than her sisters and is covered with soft, damp hairs. What a joy this is to witness. Savor the moment!

Courtesy of USDA-ARS, Stephen Ausmus

**FIGURE 8-9:** This young adult bee is just emerging from her cell.

## Watching for swarm cells

During weeks six through eight, continue looking for queen cells, but also be on the lookout for numerous queen cells in the lower third of the frames. They are swarm cells (as mentioned previously in this chapter), which are an early indication that the hive may be preparing to swarm. During your first season, don't be too concerned if you spot an occasional swarm cell. It isn't likely that a new colony will swarm. However, when you find eight or more of these swarm cells, you can be fairly certain the colony intends to swarm.

**WARNING**

You don't want a swarm to happen. When a colony swarms, half the population leaves with the old queen looking for a new and more spacious home. Before that happens, the bees take steps — evidenced by the presence of swarm cells — to create a new queen. But with half of the girls gone, and several weeks lost while the new virgin queen gets up to speed, you're left with far fewer bees gathering honey for you. Your harvest will be only a fraction of what it might otherwise have been. Prevent this unhappy situation before it happens by anticipating the bees' need for more space and adequate ventilation. (See Chapter 10 for more about swarming.)

## Providing more ventilation

During weeks five or six, you need to improve hive ventilation by opening the hive entrance. Turn the entrance reducer so the larger of its two openings is in position. The larger opening is about 4 inches wide. (Eventually, the reducer is completely removed.) The colony is now robust enough to protect itself, and the weather is milder.

You can remove the entrance reducer completely in the eighth week following the installation of your bees.

## Manipulating the frames of foundation

By the seventh or eighth week, manipulate the order of frames in a Langstroth hive to encourage the bees to draw out more foundation into comb cells. You can do this by placing any frames of foundation that haven't been drawn between frames of newly drawn comb. However, don't place these frames smack in the middle of the brood nest. That would be counterproductive because doing so splits (or breaks) the nest apart, making it difficult for bees to regulate the environment of the temperature-sensitive brood.

## Making room for honey!

As the eighth week approaches, you may find that the bees, in addition to drawing seven or more frames in the lowest box, may have drawn out seven of the ten frames in the upper deep of your Langstroth hive. When that happens, remove the hive-top feeder (if you're using one) and add a queen excluder and a shallow honey super with frames and foundation (see Chapter 4 for more information on woodenware). The girls now are ready to start collecting honey for you!

The act of adding shallow (honey) supers to a colony is called *supering*.

## WHAT TO DO ABOUT PROPOLIS DURING INSPECTIONS

Here's a question I'm often asked: "My bees have built comb on and around some of the frames. There's comb along the bottom bars, and they've glued some of the frames together with propolis. Should I scrape this wax and propolis off, or should I leave things as the bees intended?"

With every inspection you should take a moment to scrape extra wax and propolis off the wooden parts. Don't let it build up, or the job will become too daunting to do anything about it. Take a few minutes during every inspection to tidy things up. Getting into the habit of cleaning house will save you tons of work later on. If you allow the bees to glue everything together, it's tempting to forgo a necessary inspection to avoid the challenge of pulling things apart. Don't let the bees get the upper hand. Scrape it off!

## Inspecting your multilevel Langstroth hive

Now that there are two hive bodies on your Langstroth hive (later there will also be honey supers), you have more than just one box to inspect. In your first season, to gain experience you should inspect the entire colony: both the lower deep and the upper deep. The process is the same as any inspection, except you will first need to systematically remove all the honey supers and the upper deep to get down to the lower deep. That's where you will begin your frame-by-frame inspection. In subsequent years, after gaining experience, you need not inspect every frame. At that point, you can evaluate your colony by inspecting only a sampling of frames.

Once all the frames in the lower deep are inspected, replace the upper deep and begin your frame-by-frame inspection of it. As always, you are checking for evidence of the queen, a good laying pattern, and healthy brood.

As the season progresses, the upper deep will be used by the bees to store honey, so eventually you won't find the queen, any eggs, or brood in the upper deep. The lower deep is where the action is as summer progresses. That will be the focus of your inspection.

**TIP**

Use the outer cover (placed bottom-side-up on the ground) as a surface on which to place supers and hive bodies as you disassemble the hive for inspection (see Figure 8-10).

**FIGURE 8-10:**
Use the outer cover (placed on the ground bottom-side-up) as a surface on which to place supers and hive bodies as you disassemble the hive for inspection.

*Courtesy of Howland Blackiston*

# Chapter **9**

# Different Seasons, Different Activities

The seasonal calendar of events in Maine obviously looks different than one in Southern California. But different climates mean different schedules and activities for the hive and beekeeper. Regardless of their precise location, honey bees are impacted by the general change of seasons. Knowing what major activities are taking place within the hive and what's expected of you during these seasons is useful. For a good beekeeper, *anticipation* is the key to success.

This chapter contains a suggested schedule of seasonal activities for the beekeeper. However, you must note that geography, weather, climate, neighborhood, and even the type of bees influence the timing of these activities. The book is written from the viewpoint of a beekeeper experiencing a distinct change in seasons and climate (spring, summer, autumn, winter).

There is a beekeeper's calendar and map at the end of this chapter. Use this as a guide to relate the timing of these activities to where in the world you live.

The seasonal to-do lists for your Top Bar hive will be similar to those outlined for a Langstroth colony, but the manipulations can be quite different. As you go through this chapter, look for tips and other helpful details that highlight the differences.

I also suggest some important tasks for the beekeeper and provide a rough estimate of the amount of time you'll need to spend with your bees during each season. These time estimates are based on maintaining one to three hives.

# Lazy, Hazy, Crazy Days of Summer

Nectar flow usually reaches its peak during summer. That's also when the population of the colony usually reaches its peak. When that's the case, your colonies are quite self-sufficient, boiling with worker bees tirelessly collecting pollen, gathering nectar to convert into honey, and building beeswax comb to store both. Note, however, that the queen's rate of egg laying drops during the late summer. And on hot and humid nights, you may see a huge curtain of bees hanging on the exterior of the hive. Don't worry; they're not running away. They're simply cooling off on the front porch. It's called *bearding* (see Figure 9-1).

**FIGURE 9-1:** On this summer evening, the bees were found cooling off on the exterior of my hive.

*Courtesy of Howland Blackiston*

Late in summer the colony's growth begins to diminish. Drones still are around, but outside activity begins slowing down when the nectar flow slows. Bees seem to be restless and become protective of their honey.

# Your summer to-do list

Here are some activities you can expect to schedule between trips to the beach and hot-dog picnics:

>> Inspect the hive every few weeks; make sure it's healthy and the queen is present.

>> Add honey supers as needed. Keep your fingers crossed in anticipation of a great honey harvest.

>> Keep up swarm control through midsummer (see Chapters 8 and 10). Late in the summer there is little chance of swarming.

>> Be on the lookout for honey-robbing wasps or robbing bees from other colonies. A hive under full attack is a nasty situation (see Chapter 10 for information about how to deal with robbing).

>> Harvest your honey crop at the end of the nectar flow (see Chapters 16 and 17). Remember that in zones experiencing cold winters, the colony requires at least 60 pounds of honey for use during winter.

**WARNING**

Honey harvest season is the time to break out your gloves, because your normally docile bees are at their most defensive. They don't want to give up their honey without a bit of a fight!

**TIP**

A Top Bar hive colony will sometimes build honeycomb on both sides of the brood nest. This behavior can lead to the colony becoming "honey bound" with no place for the queen to lay. As you inspect your Top Bar hive colony and notice this behavior, you can simply move the honeycombs toward the rear and insert a few blank bars. This will allow the brood to continue to expand and also help control the urge to swarm.

## Your summertime commitment

You can't do all that much until the end of the summer and the honey harvest because your bees are doing it all! Figure on spending about eight to ten hours with your bees during the summer months. Most of this time involves harvesting and bottling honey (see Chapters 16 and 17 for more information on honey harvesting).

# Falling Leaves Point to Autumn Chores

Most nectar and pollen sources become scarce as days become shorter and weather cools in autumn. All in all, as the season slows down, so do the activities within your hive: The queen's egg-laying is dramatically reduced, drones begin to disappear from the hive, and hive populations drop significantly.

Your bees begin bringing in propolis, using it to chink up cracks in the hive that may leak the winter's cold wind. The colony is hunkering down for the winter, so you must help your bees get ready.

## Your autumn to-do list

When helping your bees prepare for the upcoming hardships of winter months, you must

>> Inspect your bees (look inside the hive) and make certain that the queen is there. As mentioned in Chapter 8, the easiest way to know you have a queen is when you find eggs. One egg per cell means the queen is present.

Be sure to look for eggs, *not* larvae. Finding eggs means that the queen was present within the last three days. Larvae, on the other hand, can be three to eight days old. Thus, merely finding larvae is no guarantee that you still have a queen.

When you wait too late during autumn, you discover that eggs and larvae are few and far between. In that case, actually finding the queen is the surest way to check. Be patient and look carefully.

>> Determine whether the bees have enough honey. Your bees need plenty of food (capped honey) for the winter. Make certain that the upper deep hive body is full or nearly full of honey. Honey is essential for your bees' survival, because it's the fuel that stokes their stoves. Without it, they're certain to perish.

**REMEMBER**

In cooler, northern climates, hives need about 60 pounds or more of honey as they head into winter. You'll need less honey reserves (30 to 40 pounds) if your winters are short (or nonexistent).

>> Feed your colony. They'll accept a 2-to-1 sugar-syrup feeding (see the "Autumn syrup recipe" sidebar) until colder weather contracts them into a tight cluster. At that point, temperatures are too cold for them to leave the cluster (see the "Clustering in a Winter Wonderland" section later in this chapter), so they are likely to ignore your offering.

**TIP**

The amount of honey needed for a Top Bar colony to survive winter depends on the climate where the colony is kept. In northern climates, a Top Bar colony can require about ten solid combs of honey, which is equivalent to about 60 pounds, depending on the width and depth of your particular hive. Another consideration is the depth of the honey band on each bar (see Figure 9-2). The honey band forms above the brood similar to what you find on a typical brood frame in a Langstroth box. Thin bands of honey on the brood comb can be a problem for the winter cluster since the winter cluster always starts on those bars. If a situation develops where the cluster is unable to break up and move to solid honey, the colony can starve with thin bands. If your honey bands are less than 2 inches wide on the brood bars, feeding your colony to increase the band's width is important.

**FIGURE 9-2:**
The proper honey band width in a Top Bar hive to accommodate a winter cluster until the bees can break cluster and move. Less than this amount in a Top Bar frame might prove fatal.

*Courtesy of William Hesbach*

>> Keep feeding your bees until they stop taking the syrup, or until the temperature drops, and they form the winter cluster. A hive-top feeder works best. If you are medicating, the first 2 gallons should be medicated with Fumigilin-B — subsequent feedings are not medicated.

>> Provide adequate ventilation. During winter, the temperature at the center of the cluster is maintained at 90 to 93 degrees Fahrenheit (34 to 35 degrees Celsius). Without adequate ventilation, the warm air from the cluster rises and hits the cold inner cover, and condensation drips down onto the bees as

ice-cold water. That's a big problem! The bees will become chilled and die. Keep your colony dry by doing the following:

- Glue (permanently) four postage stamp–sized pieces of wood (you can use the thin end of a wood shingle or pieces of a Popsicle stick) to the four corners of the inner cover's flat underside. This neat ventilation trick makes an air space of $\frac{1}{16}$ inch or less between the top edge of the upper deep hive body and the inner cover.

- If you are using a Langstroth hive, place the inner cover on the top deep body, *flat side down*. The oval hole (if there is a hole) should be left open, and the ventilation notch cut into the ledge of the inner cover (if there is one) should be left open, too.

**REMEMBER**

When you put the outer cover on a Langstroth hive, make sure you push it forward so the notch in the ledge of the inner cover remains open. Make sure the outer cover is put on the hive equidistant from side to side. The result is a gentle flow of air that carries off moisture from the underside of the inner cover and thus keeps the colony dry.

## AUTUMN SYRUP RECIPE

I use special "heavy" syrup for feeding bees that are about to go into the winter months. The thicker-consistency recipe makes it easier for bees to convert the syrup into the capped honey they'll store for the winter.

Boil 2½ quarts of water on the stove. When it comes to a rolling boil, turn off the heat and add 10 pounds of white granulated sugar. Be sure you turn off the stove. If you continue boiling the sugar, it may caramelize, and that makes the bees sick. Stir until the sugar dissolves completely. The syrup must cool to room temperature before you can add medication. Note that this is a recipe for thicker syrup than that used for feeding bees in the spring.

If you see clear evidence of Nosema (you'll notice mustard brown drips on the hive and at the entrance), you may want to medicate your syrup. A colony with Nosema will have a hard time surviving the winter. (See Chapter 12 for more information on Nosema.) To medicate the syrup, mix 1 teaspoon of Fumigilin-B in approximately a half cup of luke-warm water (the medication won't dissolve in the syrup). Add the medication to the syrup and stir. Only the first 2 gallons you feed your bees need to be medicated. Subsequent batches do not need to be medicated.

**TIP**

- A good way to ventilate a Top Bar hive is to leave a small space between a few bars in the front near the entrance. By small I mean less than a bee-space (⅜ inch). How much ventilation you need is a matter of discovery based on the physical location of your apiary and your regional climate. One indication that some ventilation is required is excessive moisture on the hive sides and top, and the face of the combs. Excess moisture also supports the growth of mold on combs. During your inspections at the end of the season, be observant and note how moisture is building up. Then open some top bar space as required.

» Wrap the hive in black tar paper (the kind used by roofers; see Figure 9-2) if you're in a climate where the winter gets below freezing for more than several weeks. Make sure you don't cover the entrance or any upper ventilation holes. The black tar paper absorbs heat from the winter sun and helps the colony better regulate temperatures during cold spells. It also acts as a windbreak. This kind of wrap can be done on either a Langstroth hive or a Top Bar hive.

**TIP**

I put a double thickness of tar paper over the top of the hive. Placing a rock on top ensures that cold winds don't lift the tar paper off. I also cut a hole in the wrapping to accommodate the ventilation hole I drilled in the upper deep hive body (see Figure 9-3).

» Provide a windbreak if your winter weather is harsh. Hopefully, you originally were able to locate your hives with a natural windbreak of shrubbery (see Chapter 3). But if not, you can erect a temporary windbreak of fence posts and burlap. Position it to block prevailing winter winds.

## MAKING YOUR WINTER VENTILATION PREPARATIONS A BREEZE

Here's an easy ventilation trick from a commercial beekeeper who has successfully overwintered thousands of hives in upper New York State. During your late-autumn preparation, simply slide the Langstroth's upper deep back so you create a ⅛-inch opening along the entire front lower hive body. Don't make it a larger gap, or the bees will use it as an entrance or you might create a robbing situation. "Wait a minute!" you might say. "Doesn't the rain get in that little gap?" Yes, it does. But that's no problem because you've already tilted your hive slightly forward (see Chapter 3). Any rain or snow that dribbles in simply drains right out the front door. Try this trick along with your other ventilation routines. But wait until the weather turns chilly before you do this, or the bees will use propolis to close the gap.

**FIGURE 9-3:**
Wrap your hive in tar paper to protect your colony from harsh winter winds and absorb the warmth of the sun. The rock on top keeps the paper from blowing off; the metal mouse guard keeps unwanted visitors out; and the hole in the upper deep aids ventilation.

*Courtesy of Howland Blackiston*

**TIP**

If you want a little extra protection for your Top Bar hive, a foam covering is an alternative to tarpaper (see Figure 9-4). Foam provides your Top Bar hive with excellent insulation in addition to a windbreak. You can easily fabricate the panels from material purchased at a home center. The most important surface to insulate is the top of the bars because most heat is lost from the top. You can simply cut a thin, flat piece that fits inside your outer cover, and you're done.

>> Add a mouse guard to the front entrance of the hive (see Chapter 13 for more information on mouse guards).

**FIGURE 9-4:**
A foam covering serves as insulation and a windbreak for your Top Bar hive.

*Courtesy of William Hesbach*

## Your autumn time commitment

Figure on spending three to five hours total to get your bees fed and bedded down for the winter months ahead.

# Clustering in a Winter Wonderland

A lot goes on inside the hive during the winter. The queen is surrounded by thousands of her workers — kept warm in the midst of the winter cluster. The winter cluster starts in the brood chamber when ambient temperatures reach 54 to 57 degrees (12 to 14 degrees Celsius). When cold weather comes, the cluster forms in the center of the two hive bodies. It covers the *top* bars of the frames in the lower chamber and extends over and beyond the *bottom* bars of the frames in the food chamber (see Figure 9-5).

**FIGURE 9-5:** Although the outside temperature may be freezing, the center of the winter cluster remains toasty warm. This cutaway illustration shows the winter cluster's position.

*Courtesy of Howland Blackiston*

Although the temperature outside may be freezing, the center of the winter cluster remains between 90 and 93 degrees (32 and 34 degrees Celsius). The bees generate heat by "shivering" their wing muscles.

No drones are in the hive during winter, but some worker brood begin appearing late in the winter. Meanwhile, the bees consume about 50 to 60 pounds of honey in the hive during winter months. They eat while they are in the cluster, moving around as a group whenever the temperature gets above 40 to 45 degrees (4 to 7 degrees Celsius). They can move to a new area of honey only when the weather is warm enough for them to break cluster.

Bees won't defecate in the hive. Instead, they hold off until they can leave the hive on a nice, mild day when the temperature reaches 45 to 50 degrees (7 to 10 degrees Celsius) to take *cleansing flights.*

## Your winter to-do list

Winter is the slowest season of your beekeeping cycle. You've already prepared your colony for the kinds of weather that your part of the world typically experiences. So now is the time to do the following:

>> Monitor the hive entrance. Gently brush off any dead bees or snow that blocks the entrance. (You don't want to disturb the bees inside.)

>> Make sure the bees have enough food! The late winter and early spring are especially hazardous because during this time colonies can die of starvation.

Late in the winter, on a nice, mild day when there is no wind and bees are flying, take a quick peek inside your hive. It's best not to remove any frames. Just have a look-see under the cover. Do you see bees? They still should be in a cluster in the upper deep. Are they okay?

If you don't notice any sealed honey in the top frames, you may need to begin some emergency feeding. But remember that once you start feeding, you *cannot* stop until the bees are bringing in their own pollen and nectar.

>> Clean, repair, and store your equipment for the winter.

>> Attend bee club meetings and read all those back issues of your favorite bee journals.

>> Order package bees and equipment (if needed) from a reputable supplier.

>> Try a bee-related hobby. The winter is a good time for making beeswax candles, brewing some mead, and dreaming of spring! See Chapter 18 for some fun ideas.

## Your winter time commitment

During this time, the bees are in their winter cluster, warm inside the hive. Figure on spending two to three hours repairing stored equipment, plus whatever time you may spend on bee-related hobbies — making candles, mead, cosmetics, and so on — or attending bee-club meetings. You might even decorate your hive for the holidays.

# Spring Is in the Air (Starting Your Second Season)

Spring is one of the busiest times of the year for bees (and beekeepers). It's the season when new colonies are started and established colonies come back to life.

Days are getting longer and milder, and the established hive comes alive, exploding in population. The queen steadily lays more and more eggs, ultimately reaching her greatest rate of egg-laying. The drones begin reappearing, and hive activity starts hopping. The nectar and pollen begin coming into the hive thick and fast. The hive boils with activity.

Most of the information that follows applies to both Langstroth hives and Top Bar hives. I've included a separate section entitled "Managing Top Bar Hives in the Spring" at the end of this chapter that highlights a few differences.

## Your spring to-do list

Beekeepers face many chores in the springtime, evaluating the status of their colonies and helping their bees get into shape for summer months. Some of those chores include the following:

**TIP**

>> Conduct an early-bird inspection. Colonies should be given a quick inspection as early in the spring as possible. The exact timing depends on your location (earlier in warmer zones, later in colder zones).

You don't need to wait until bees are flying freely every day nor until the signs of spring are visible (the appearance of buds and flowers). Do your first spring inspection on a sunny, mild day with *no wind* and a temperature close to 50 degrees (10 degrees Celsius).

A rule of thumb: If the weather is cold enough that you need a heavy overcoat, it's too cold to inspect the bees.

>> Determine that your bees made it through the winter. Do you see the cluster? The clustered bees should be fairly high in the upper deep hive body. If you don't see them, can you hear the cluster? Tap the side of the hive, put your ear against it, and listen for a hum or buzzing.

If it appears that you've lost your bees, take the hive apart and clean out any dead bees. Reassemble it and order a package of bees as soon as possible. Don't give up. We all lose our bees at one time or another.

>> Check to make sure you have a queen. If your hive has a live colony, look down between some of the frames. Do you see any brood? That's a good sign that the queen is present. To get a better look, you may need to carefully remove a frame from the center of the top deep. Can you see any brood? Do you spot any eggs?

**WARNING**

This inspection must be done quickly because you don't want to leave the frame open to chilly air. If you don't see any brood or eggs, your hive may be without a queen, and you should order a new queen as soon as possible, assuming, that is, the hive population is sufficient to incubate brood once the new queen arrives. What's sufficient? The cluster of bees needs to be *at least* the size of a large grapefruit (hopefully larger). If you have fewer bees than that, you should plan to order a new package of bees (with queen).

>> Check to ensure the bees still have food. Looking down between the frames, see if you spot any honey. Honey is capped with white cappings (tan cappings are the brood). If you see honey, that's great. If not, you must begin emergency feeding your bees (see the following bulleted items).

>> Feed the colony. A few weeks before the first blossoms appear, you need to begin feeding your bees (regardless of whether they still have honey).

Feed the colony sugar syrup (see recipe in Chapter 6). This feeding stimulates the queen and encourages her to start laying eggs at a brisk rate. It also stimulates the worker bees' wax glands. Continue feeding until you notice that the bees are bringing in their own food. You'll know when you see nuggets of pollen on their legs.

**TIP**

Feed the colony pollen substitute, which helps strengthen your hive and stimulates egg-laying in the queen. Pollen substitute is available in a powdered mix or already formed into a ready-to-feed patty from your bee supplier. This feeding can cease when flight conditions improve and you see bees bringing in their own pollen.

>> Reverse your hive bodies. See the "Reversing hive bodies" section later in this chapter.

>> Anticipate colony growth. Don't wait until your hive is "boiling" with bees. Later in the spring, before the colony becomes too crowded, create more room for the bees by adding a queen excluder and honey supers. Be sure you remove the feeder and discontinue all feeding at this time.

>> Watch out for indications of swarming. Inspect the hive periodically and look for *swarm* cells (see Chapter 8).

# Your springtime commitment

Spring is just about the busiest time for the beekeeper. You can anticipate spending 8 to 12 hours tending to your bees.

# Administering spring medication

**ALL NATURAL**

Those practicing a *natural* approach to beekeeping may want to avoid medication treatments. It's up to you what kind of beekeeping you plan to practice. I've included this section to help those beekeepers who think it prudent to medicate their bees.

You probably don't need to medicate your bees during their first season (reputable bee suppliers should provide you with strong and healthy bees). Remember to play it safe and stop all medication treatments a couple of weeks *before* adding honey supers to the colony to prevent contamination of the honey that you want to harvest.

The list that follows contains a springtime medication regime that helps treat suspected diseases and control mites (see Chapters 11, 12, and 13 for more information):

>> **Control Nosema (if evident):** In a small jar half filled with lukewarm water, add 1 teaspoon of *Fumigilin-B*. Shake the jar until dissolved. Stir the jar's contents into the cooled sugar syrup solution you use to feed your bees (see Chapter 6 for a recipe). Feed at the top of the hive using a hive-top feeder. Medicate the first 2 gallons of syrup, but not subsequent gallons.

>> **Prevent foulbrood:** A number of the antibiotics previously available to deal with foulbrood have either been taken off the market or may soon be taken off the market. This is the result of their overuse and the resulting increase in strains of foulbrood now resistant to those antibiotics. One brand that is still on the market as of this writing is TetraBee (containing the antibiotic oxytetracycline). This powder is a preventative against American and European foulbrood. To administer, carefully follow the directions on the package.

>> **Control Varroa mites:** There are a number of choices for treatment (see Chapter 13 for more information on Varroa mite control). If using a medication approach, it's important that you follow package directions precisely.

**WARNING**

Never, ever leave medicated mite control products in the hive over the winter. Doing so constantly exposes the mites to the active chemical ingredient, which becomes weaker and weaker over time. These sublethal doses increase the chance for mites to build up a resistance to the products. This tolerance is then passed on to future generations of mites, and subsequent treatments become less and less useful.

**ALL NATURAL**

>> **Control tracheal mites:** When the weather starts getting warmer, place a prepared bag containing 1.8 ounces of *menthol crystals* on the top bars toward the rear of the hive (see Chapter 13 for more information). Set them on a small piece of aluminum foil to prevent the bees from chewing holes in the bag and carrying it away. Leave the bag in the hive for 14 *consecutive* days when the outdoor temperature ranges between 60 and 80 degrees (16 and 27 degrees Celsius).

Note that many consider the use of menthol a *natural* approach to dealing with tracheal mites. Menthol is an organic compound made synthetically or obtained from cornmint, peppermint, or other mint oils.

Adding a grease patty to the top bars of the brood chamber is another natural treatment for tracheal mites. Making grease patties is easy; see Chapter 13 for my favorite recipe. Use one patty per hive, replacing them as the bees consume them. Remove these patties when honey supers are on the hive. Unused patties can be stored in the freezer until you're ready to use them.

## Reversing hive bodies

Bees normally move upward in the Langstroth hive during the winter. In early spring, the upper deep is full of bees, new brood, and food. But the lower deep hive body is mostly empty. You can help colony expansion by reversing the top and bottom deep hive bodies (see Figure 9-6). Doing so also gives you an opportunity to clean the bottom board. Follow these steps:

1. **When a mild day comes along (50 degrees [10 degrees Celsius]) with little or no wind and bright, clear sunlight, open your hive using your smoker in the usual way.**

**FIGURE 9-6:** Reversing hive bodies in the spring helps to better distribute brood and food and speeds up the growth of your colony's population.

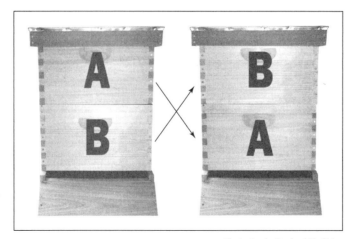

*Illustration by Howland Blackiston*

2. **Place the upturned outer cover on the ground and then remove the upper deep hive body with the inner cover in place.**

3. **Place the deep across the edges of the outer cover so there are only four points of contact (you'll squeeze fewer bees that way).**

4. **Close the oval hole in the middle of the inner cover with a piece of wood shingle or tape.**

5. **Now you can see down into the lower deep that still rests on the bottom board.**

   It probably is empty, but even if some inhabitants are found, lift the lower deep off the bottom board and place it crossways on the inner cover that is covering the deep you previously removed.

6. **Scrape and clean the bottom board.**

   *Note:* This is a good opportunity to add a slatted rack (see Chapter 5). Slatted racks help with the hive's ventilation and can promote superior brood patterns. They also encourage the queen to lay eggs all the way to the front of the hive because of improved ventilation and draft control.

7. **Now stand the deep body — which had been the relatively empty bottom one — on one end, placing it on the ground.**

8. **Place the *full* hive body (the one that had been on top) onto the clean bottom board (or on the slatted rack, if you added one).**

9. **Smoke the bees and remove the inner cover so you can place the empty deep (the box that had been the lowest) on top.**

10. **Replace the inner and outer covers.**

This reversing procedure enables the bees to better distribute brood, honey, pollen, fresh nectar, and water. Reversing gives them more room to move upward, which is the direction they always want to move.

Repeat this reversal in about three to four weeks, restoring the hive to its original configuration. At that time you can add one or more honey supers above the two brood boxes — assuming the bees are now bringing in their own food and you have ceased feeding.

# Managing Top Bar Hives in the Spring

After successfully wintering your Top Bar hive, the colony comes into early spring with vigor. The queen has been laying in small patches since early January, and the colony is prepared to build a large foraging population that will be ready when the main nectar flow starts. Make certain that the colony is "queen right" and the drive to reproduce (and later swarm) is monitored and managed.

## Finding the cluster

One issue you will need to determine is the location of the cluster and its relative size. A Top Bar hive affords a quick and easy way to determine both without the need to open the colony and can be done quickly at any time of year. You simply remove the cover and slide your bare hand across the top bars, searching for a temperature change. As discussed earlier, the winter cluster, or for that matter any cluster, will be in the 90-degree (30-degree Celsius) range when actively rearing brood, and the bars directly over the cluster will be noticeably warmer.

Sensing the top-bar temperatures will enable you to outline the size and location of the cluster. Knowing where the cluster is will direct your inspections to that specific location and allow you to determine how to enter the colony without much disturbance of the bees. In a northern climate, an adequate spring cluster is five or six bars.

## Preventing the urge to swarm

Your second-year colony will most likely initiate a reproductive swarm (Chapter 10 explains swarming). In a Top Bar hive, the swarm cells are likely to be located on the comb's edge (see Figure 9-7).

**FIGURE 9-7:** When this newly forming swarm cell is capped, the colony will likely swarm. Notice the care the nurse bees are giving to the young larva.

*Courtesy of William Hesbach*

So what to do about a Top Bar hive colony building swarm cells? One option is to have extra equipment on hand so you can split the colony in half by moving some frames of open brood to other equipment. The new colony will then raise a new queen. This should end the urge to swarm. Another option is to allow the colony to swarm with the intention to capture it yourself. But with this option you might miss capturing the swarm and lose your beautiful queen and, chances are, the colony will not make much honey that season. It's your choice when it comes to swarm management. I suggest you experiment with both options and use this experience to learn more about the reproductive biology of bees.

## Expanding the brood nest

Springtime is all about colony expansion. Your Top Bar hive will begin to build out brood in the direction of the entrance. With incoming nectar and pollen, your job will be to ensure that the brood nest has plenty of room to expand by supplying new bars as needed or by moving the follower board so you always have two to three empty top bars waiting for bee expansion. Observe what your colony is doing and insert either blank bars or more drawn comb from the previous season. If the colony feels it has run out of room to raise brood, it may decide to initiate a swarm.

# The Beekeeper's Calendar

This is not the bee-all and end-all of a to-do list! It's simply a guideline to help you determine the kind of chores you should consider as the season progresses. The activities apply to both Langstroth and Top Bar hives. Note that a beekeeper's calendar of activities will be different in Maine than in Southern California (see Figure 9-8). And the corresponding dates and activities can vary depending upon actual weather conditions, elevation, and so on. Consider this tool a "sanity check" as you and your bees progress through the seasons.

**Zone A:** Short summers and long, cold winters. Average annual temperature is between 35 and 45 degrees (2 and 7 degrees Celsius). Minimum temperatures are between 0 and 15 degrees (−18 and −9 degrees Celsius).

**Zone B:** Summers are hot, and winters can be quite cold and extended. Average annual temperature is between 45 and 55 degrees (7 and 13 degrees Celsius). Minimum temperatures are between 15 and 20 degrees (−9 and −7 degrees Celsius).

**Zone C:** Summers are long and hot, and the winters are mild and short. Average annual temperature is between 55 and 65 degrees (13 and 18 degrees Celsius). Minimum temperatures are between 30 and 35 degrees (−1 and 2 degrees Celsius).

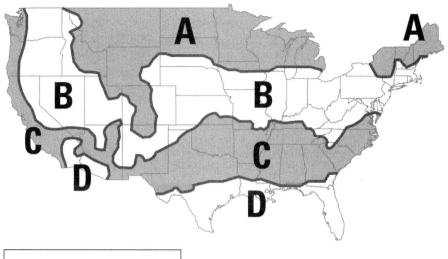

**FIGURE 9-8:**
Beekeeping
temperature
zones in the
United States.

*Illustration by Howland Blackiston*

**Zone D:** Warm to hot all year round. Average annual temperature is between 65 and 80 degrees (18 and 27 degrees Celsius). Minimum temperatures are between 30 and 40 degrees (−1 and 4 degrees Celsius).

Following are some guidelines on how to use this tool:

1. Use Figure 9-8 to determine your beekeeping temperature zone. If you live outside the United States, find the zone on the map with a temperature range that most closely corresponds to your part of the world.

2. Use Table 9-1 to locate the month of year you are currently in.

3. Look down the specific month column and find your zone letter (A, B, C, or D). Wherever your zone letter appears, look at the corresponding activity in the far-left column. This is an activity you should consider doing during this month. All of these activities are covered in more detail within the various chapters of the book.

**TABLE 9-1    Beekeeping Calendar**

| Typical Activity | Jan | Feb | Mar | Apr | May | Jun | Jul | Aug | Sep | Oct | Nov | Dec |
|---|---|---|---|---|---|---|---|---|---|---|---|---|
| Check food reserves | B | B | A, C | A | A | | B | B | B | A, C, D | | |
| Feed colony if low on capped honey | B, D | B, D | A, B | A, B | A | | B | B | B | A, B, C, D | C, D | D |
| Check for eggs/queen | | D | B, C | B, C | A | A | | | C | D | | |
| Reverse hive bodies | | | B | C | A | | | | | | | |
| Install new bees in hive | | | | B, C, D | A, B | | | | | | | |
| First comprehensive inspection of the season | | C, D | C, B | B | A | | | | | | | |
| Check for capped brood and brood pattern | | D | B | B, C | A, B, C | A, B | A, B | B | | | | |
| Feed a pollen substitute | | C, D | A, B, C | A, B, C | | | | | | | | |
| Look for swarm cells | | | D | B, D | A, B, C, D | A, B, C | | | | | | |
| Add queen excluder and honey supers | | | D | B, D | A, B, C, D | A | | | | | | |
| Look for supercedure cells | | | | B, C, D | A, B, C, D | A, B, C, D | | | | | | |
| Check ventilation | | | | | | D | B, C, D | C, D | B | A | A | |
| Add mouse guard | | | | | | | | | B, C | A, B | | |
| Medicate for AFB, EFB, and Nosema | | D | C | B | A | | | A, B | C, D | | | |
| Check surplus honey | | | D | D | D | A, B, D | A, B, C | A, B | C | | | |

*(continued)*

**TABLE 9-1** (*continued*)

| Typical Activity | Jan | Feb | Mar | Apr | May | Jun | Jul | Aug | Sep | Oct | Nov | Dec |
|---|---|---|---|---|---|---|---|---|---|---|---|---|
| Harvest honey | | | | | D | B, D | B, D | B | | | | |
| Test for Varroa mites | | | C, D | B | A | D | | A, B | A, D | | | |
| Medicate for Varroa mites (if needed) | | | C, D | B | A | D | | A, B | A, C, D | | | |
| Medicate for tracheal mites | | | C, D | B | A | | | A, B | A, C, D | | | |
| Check hives for small hive beetle; medicate if needed | | | C, D | | | | | | C, D | | | |
| Prepare hive for winter | | | | | | | | | A, B | A, B, C | C | |
| Check entrance for blockage | A, B, C | A, B, C | | | | | | | | | A | A, B |
| Order new bees | A, B, C, D | | | | | | | | | | | |

# 4

# Common Problems and Simple Solutions

Find out how to keep your bees from swarming, getting sick, or undergoing stress.

Read the latest information on the highly publicized dying of honey bees as a result of colony collapse disorder, and learn what you can do.

Discover how to easily identify common bee diseases and what your options are for dealing with bee health issues.

Understand various "natural" approaches to maintaining bee health that avoid or minimize the use of chemicals in the hive.

Become familiar with the various other creatures that can negatively impact the health and well-being of your honey-bee colony. See what you can do to avert trouble before it happens.

Ensure a vigorous and healthy colony by learning how to raise your own queen bees that are bred for highly desirable genetic traits.

Chapter **10**

# Anticipating and Preventing Potential Problems

D espite the best intentions and the most careful planning, things occasionally go wrong. It happens. The bees swarm. The queen is nowhere to be found. The whole colony dies or flies away. What happened? Did you do something wrong? What could you have done differently?

I've made just about every mistake in the book at one time or another. But that's nothing to be ashamed of. It's part of the process. The key lesson I've learned has been to *anticipate*. Discipline yourself to plan ahead and look out for potential problems *before* they happen. I can assure you that you can head off 80 to 90 percent of potential problems if you anticipate trouble and take steps to avoid it.

In this chapter, I include a few of the more common nonmedical problems to anticipate and try to avoid. These problems include swarming and absconding; losing your queen; and losing your colony because of poor ventilation, robbers (robber bees, that is!), and pesticides. This chapter also tells you how to deal with potential community-mindset problems of having Africanized bees in your geographical area.

# Running Away (to Join the Circus?)

Sometimes bees disappear. They simply get up and go. Poof! In one common scenario, called *swarming,* about 50 percent of the colony packs up with the queen and takes flight. In the other scenario, called *absconding,* 100 percent of the colony hits the road, leaving not a soul behind. Neither scenario is something you want to happen.

## Swarming

A swarm of honey bees is a familiar sight in the spring and early summer. It's one of the most fascinating phenomena in nature and an instinctive way that honey bees manage the colony's growth and survival. To witness a swarm pouring out of a hive is simply thrilling — though the pleasure may be less so if the swarm of bees is yours!

Starting about a week before swarming, the bees that intend to leave the colony gorge themselves with honey (like packing a box lunch before a long trip). Then, all at once, like someone flipped a switch, tens of thousands of bees exit the hive and blacken the sky with their numbers. Half or more of the colony leaves the hive to look for a new home. But first, within a few minutes of departing from the hive, the bees settle down on a surface not too far from the hive they just left.

There's no telling where a swarm might land. It could land on any convenient resting place: a bush, a tree branch (see Figure 10-1), a lamppost, or perhaps a piece of patio furniture (see Figure 10-2). In any case, the swarming bees won't stay there long. As soon as scout bees find a more suitable and protected home, the swarm will be up, up, and away.

In its temporary resting place, the swarm is a bundle of bees clustered together for protection and warmth. Their queen is in the center of it all. Depending on the size of the hive that swarmed, the cluster may be as small as a grapefruit or as large as a watermelon. The bees will remain in this manner for a few hours or even a few days while scout bees look for a new home. When they return with news of a suitable spot, off they all go to take up residence in a hollow tree, within the walls of an old barn, or in some other cozy cavity.

**FIGURE 10-1:**
A swarm resting
in a tree.

*Courtesy of John Clayton*

**FIGURE 10-2:**
A swarm that
has taken up
temporary
residence under
a picnic table.

*Courtesy of John Clayton*

**TIP**

Not sure if your hive has swarmed? A regular inspection during the spring will reveal the situation. Know the key indicators: no eggs, fewer bees, and all the cells have only older larvae and/or capped brood. And there are queen cells present along the lower third of the comb.

## Understanding why you want to prevent swarming

Swarms are a dramatic sight and a completely natural occurrence for the bees, but swarms are not good news for you. A colony that swarms is far less likely to collect a surplus of honey. That means no honey harvest for you that year. A colony that loses 50 percent of its population and 50 percent of its honey will have a difficult time regaining its population and productivity. It also means the bees may have a tougher time making it through the cold winter months (assuming you have frosty weather).

It's unhappy enough news when your bees swarm, but the *later* in the season they do it, the worse the news is for you. If the bees choose to swarm later, and you live in an area that experiences cold winter months, there simply isn't enough time for the colony to recover during that season.

If you're a first-year beekeeper, rest assured that a new colony is unlikely to swarm during its first season. But older and more crowded colonies are likely candidates for swarming behavior. Remember, swarming is a natural and normal instinct for bees. At one point or another, your bees will want to swarm. It's only natural. It's nature's way of reproduction. But discouraging them from doing so is a skill every beekeeper should have because a swarm means fewer bees to make honey for you.

## Keeping the girls from leaving home

There are two primary reasons bees swarm: congestion of the brood area and poor ventilation. Occasionally, a poorly performing queen can contribute to the swarming impulse. But all these conditions can be anticipated and avoided. Here are some things you can do:

» **Avoid congestion.** Because overcrowding is a primary reason a colony will swarm, make sure to anticipate your bees' needs and provide them with more room *before* they need it. If you wait until it's obvious that the brood area is crowded, you're too late! The colony is likely to swarm, and there is little you can do to prevent them from swarming once they've set their minds to it. However, you can do the following to prevent congestion from happening in the first place:

- Reverse your hive bodies in the early spring to better distribute the fast-growing population (see Chapter 9).

- Add a queen excluder and honey supers *before* the first nectar flow in the early spring (stop feeding before you add honey supers; see Chapter 9).

» **Provide adequate ventilation.** To ensure proper ventilation, you can do a number of things:

- If your inner cover has a notched ventilation hole in the front of the inner cover, make sure it's open. Here's how. Stand at the rear of the hive and push the outer cover forward. Doing so prevents the overhang of the outer cover from blocking the notched hole on the inner cover.

- Glue a short length of a wooden Popsicle stick to each of the four corners of the inner cover. By doing so, you create a thin gap between the inner cover and the hive and improve airflow into and out of the hive. (Alternatively, you can place a short screw with a fat, domed head in each corner. The fat head of the screw creates the gap you want.)

- Drill wine cork–sized holes in your upper deep (below the hand hold) and in all your honey supers, as shown in Figure 10-3. Doing so not only provides extra ventilation but also provides the bees with additional entrances. This ventilation can even be helpful in the cold winter months. You can control airflow and access by blocking and opening these holes as needed with corks or strips of duct tape. Be sure to close off these entrances for a new colony whose population is still too small to defend all these extra openings.

» **Make the bees comfortable in hot weather by doing the following:**

- Supply a nearby water source. The bees will use this water to regulate the hive's temperature. See Chapter 3 for suggestions regarding water sources.

- Shield the hive from a full day of blazing sun, particularly if you live in a blazing hot area. Locating the hive in dappled sunlight is the best solution (see Chapter 3).

**FIGURE 10-3:**
A useful way to provide a colony with ventilation is to drill wine cork–sized holes in the hive bodies and supers.

*Courtesy of Howland Blackiston*

>> **Remove queen swarm cells — all of them.** The earliest evidence that your bees are thinking about swarming is when they start to make swarm cells (see Chapter 8). During the spring and early summer, inspect your hive every week or ten days to look for swarm cells. Most of them can be found along the bottom of the frames. If you see any, remove them by cutting them out with the sharp end of your hive tool. The colony won't swarm if it doesn't have a new queen in the making.

WARNING

This technique only works if you remove 100 percent of the swarm cells. If just one cell remains behind, the colony has the green light to swarm.

>> **Replace your queen every season.** Colonies with young queens are far less likely to swarm.

TIP

If the hive is simply boiling over with bees and you failed to take any of the above precautions, there is a last-resort emergency measure. You can remove all the frames of capped brood from the hive (with bees still on the frames) and replace them with frames of foundation. *A colony will not swarm if it does not have adequate capped brood to leave behind.*

Make sure the queen is not on any of these frames. You can use these frames of bees and brood to start a *new* hive! If there are eggs on those frames, the new hive will raise a new queen. Or you can play it safe and order a new queen from your bee supplier and install her in the new hive.

## THE 70 PERCENT RULE

If you're a first-year beekeeper with a ten-frame Langstroth hive, here's a way to remember when it's time to give a new colony more room (and do so *before* it's too late):

- When seven of the ten frames in the lower deep are drawn into comb, add a second deep hive body with frames and foundation.

- When seven of the ten frames in the upper deep are drawn into comb, add a queen excluder and a honey super.

- When seven of the ten frames in the honey super are drawn into comb, add an additional honey super.

Continue providing more room in this manner, adding more space when the bees have drawn out 70 percent of the foundation.

If you are using an eight-frame Langstroth hive, take these same actions when about five of the eight frames are drawn out into comb.

## They swarmed anyway. Now what?

Okay, the bees swarmed anyway. You're not alone; it happens. The good news is that you may be able to capture your swarm and start another colony. (See the following section titled "Capturing a swarm.") You wanted a new hive of bees anyway, didn't you?

In any event, what should you do with the half of the colony that remains? Follow these steps:

1. **A week after your colony swarms, inspect the hive to determine whether it has a new queen.**

   You may spot a queen cell or two along the lower third of the frames (see Chapter 8 for tips on finding queen cells). Good! That's an encouraging sign. It means a new queen is "in the oven." But you must ultimately determine if the colony's new queen is laying eggs. One week after a swarm you're unlikely to see any eggs — it's too soon for the new queen to get to work. But do have a look and see if you can find her majesty. If you can, great! Close up the hive and wait another week. If you *don't* see the queen, wait a couple more days and have another look.

**TIP**

After the swarm, it will take one to seven days for the queen cell to open and a new virgin queen to emerge. Then allow a week or more for her to be ready to mate with drones. After mating, it will be another two days before she starts laying eggs. That's when you should start looking for eggs.

Consider marking your new queen once you've found her. It's common for a beekeeper to place a daub of color on the queen's thorax (back). Marking queens makes them easier to find during future inspections and verifies that the queen you see is the same one you saw during previous inspections. For information on how to do this, see Chapter 14.

2. **Two weeks after the swarm, open the hive again and look for eggs.**

Do you see eggs? If so, you have a queen, and your colony is off and running. Close things up and celebrate with a glass of mead. If there's still no sign of a queen or her eggs, wait a few days and check again. Still no eggs? Then, order a new queen from your bee supplier. Hive the replacement queen as soon as she arrives (see "Introducing a new queen to the hive" later in this chapter).

**WARNING**

If you don't follow up after a swarm, the colony can easily become queenless without you ever being aware of it. No queen, no brood. No brood, no good.

1. **Prepare a new hive.**

Have at-the-ready a new hive body with nine frames and foundation, a bottom board, a hive-top feeder, and an outer cover (I'll refer to this as the new hive).

2. **Smoke the old hive and remove the frame with the queen.**

Turn your attention to the suspect hive (I'll refer to this as the old hive). Smoke and inspect, looking for the frame with the queen on it. When you find that regal frame, gently put it aside. Be careful! The queen is on that frame! You can make use of an empty nuc box or another empty hive body to hold this frame out of harm's way. In any event, find a way to keep the queen and frame safe and sound while you tend to other things.

## USING AN ARTIFICIAL SWARM TO PREVENT A NATURAL SWARM

There's another way to prevent a crowded hive from swarming: by creating an "artificial swarm" (sometimes called a "shook swarm"). This little trick is a lot of work, but it's an effective way to get the urge to swarm out of your colony's system. The best time to do a "shook swarm" is before 10 a.m. or after 2 p.m. Note that for this method to work, your "suspect" hive must have at least one queen cell on the frames.

3. **Move the old hive at least 10 feet away from its original location.**

   (Here's where a wheelbarrow or hive lifter comes in handy.)

4. **Now place the new hive setup where the old hive was previously located.**

5. **Place a bedsheet in front of the new hive, from the ground to the entrance board.**

   You are creating a ramp for the bees that you are about to unceremoniously dump in front of this hive.

6. **Transfer the majority of the bees to the new hive.**

   Back to the old hive. One by one, take each frame out of the old hive, and shake 80 to 90 percent of the bees off the frames (use a bee brush if you prefer) and onto the bedsheet ramp in front of the new hive. They will march their way right into the new hive. Make sure you don't shake all the bees off the frames. About 10 to 20 percent of the bees should remain on the old frames.

7. **Put the old frames (with some bees still clinging to them) back into the old hive.**

   At this point, the old hive has nine of its original frames containing brood, larvae, eggs, and about 10 to 20 percent of the bees. Remember that these frames must contain at least one queen cell. Add a new frame and foundation to take up the empty (tenth) slot.

8. **Take the frame with the old queen and gently brush her onto the entrance of the new hive.**

   Bee careful!

9. **Take the frame that the queen was on and slip it into the tenth slot of the new hive.**

   Your new hive now contains this "old" frame, nine new frames with foundation, and about 80 to 90 percent of the bees. Plus the original queen.

10. **Feed syrup to both hives using hive-top feeders or some other suitable feeding device.**

It's a good practice to close up the new hive for a day or two by pushing screening along the entrance way. Confining the bees in this manner gets them working on building new comb and helps them get over the swarming instinct. Be certain to remove the screening after a day or two.

## Capturing a swarm

If your bees do swarm and you can see where they landed (and you can reach it safely), you can capture them and start a new hive. You may even be lucky enough to get a call from a friend or neighbor who has spotted a wild swarm in his yard (beekeepers are often called to come capture swarms). Either way, capturing a swarm is a thrilling experience.

Despite their rather awesome appearance, swarms are not that dangerous. That's because honey bees are defensive only in the vicinity of their nest. They need this defensive behavior to protect their brood and food supply. But a swarm of honey bees has neither young nor food to defend and is usually very gentle. That's good news for you because it makes your job easy if you want to capture a swarm of bees.

WARNING

If you live in an area known to have Africanized honey bees (discussed later in this chapter), you must be very cautious — a swarm might be this undesirable strain. There's no way of telling just by looking at them. If you're in doubt, don't attempt to capture a swarm — unless you are certain this swarm originated from your hive or a hive where the bees are known to be gentle.

Be prepared for a crowd of awestruck onlookers. I always draw a crowd when I capture a swarm. Everyone in your audience will be stunned as you walk up to this mass of 20,000 stingers wearing only a veil for protection. "Look" they'll gasp, "that beekeeper is in short sleeves and isn't wearing any gloves! Is he crazy?" Only you will know the secret: The bees are at their gentlest when they're in a swarming cluster. You have nothing to fear. But your bystanders will think that your bravery is supreme. To them, you are a bee charmer — or the bravest (or nuttiest) person alive!

The easiest swarms to capture are those that are accommodating enough to collect on a bush or a low tree branch — one that you can reach without climbing a ladder. Obviously, if the branch is high up in a tree, you should not attempt your first capture! Gain experience by first capturing swarms that are easy reaches. Then you can graduate to the school of acrobatic swarm collection.

Say your swarm is located on an accessible branch. Lucky you! Follow these steps to capture it:

1. **Place a suitable container on the ground below the swarm.**

   You can use a large cardboard box (my favorite), an empty beehive, or a nuc box (see Chapters 4 and 6). This container will be the swarm's temporary accommodation while you transport the bees to their new, permanent home. The container you use should be large enough to accommodate the entire cluster of bees and a hunk of the branch they are currently calling home.

2. **Get the bees off the branch.**

   One approach is to give the branch holding the bees a sudden, authoritative jolt. Doing so will dislodge the swarm, and the bees will (hopefully) fall into the container that you have placed directly under it. If this approach works, great. But it can be tricky. The swarm may miss its mark, and you may wind up with bees all over the place. In addition, this violent dislodging tests the gentle demeanor I promised!

I prefer a more precise approach that enables you to gently place (not drop) the bees into their "swarm box." This approach works if the swarm is on a branch that you can easily trim and sever from the rest of the foliage. You'll need a pair of pruning shears — a size appropriate for the job at hand — and permission from the branch owner to lop it off. Follow these steps:

1. **Study the swarm.**

   Notice how the bees are clustered on the branches. Can you spot the main branch that's holding the swarm? Are several branches holding it? Try to identify the branch (or branches) that, if severed, will allow you to gingerly walk the branch with swarm attached over to the box. In this manner, you can place the swarm in the box, not dump it.

2. **Snip away at the lesser branches while firmly holding the branch containing the mother lode with your other hand.**

   Work with the precision of a surgeon: You don't want to jolt the swarm off the branch prematurely. When you're absolutely sure that you understand which branch is holding the bees, make the decisive cut. Anticipate that the swarm will be heavier than you imagined, and be sure you have a firm grip on the branch before you make the cut. Avoid sudden jolts or drops that would knock the bees off the branch.

3. **Carefully walk the swarm (branch and all) to the empty cardboard box and place the whole deal in the box. The bees will not leave the branch as you walk, but you should walk as gingerly as if you were walking on ice.**

3. **Close up the box, tape it shut, and you're done. Whew!**

   Get it home right away because heat will build up quickly in the closed box.

I have modified a cardboard box for swarm captures. One side contains a large "window" cutout that I have fitted with mosquito screen. This window gives the captured swarm ample ventilation. Alternatively, punch some tiny holes in the box with an awl or an ice pick to provide some ventilation.

## Hiving your swarm

You can introduce your swarm into a new hive in the following manner:

1. **Decide where you want to locate your new colony.**

   Keep in mind all the factors you need to consider when making this decision (see Chapter 3).

2. **Set up a new hive in this location.**

   You'll need a bottom board, a deep hive body, ten frames and foundation, an inner cover, an outer cover, and a hive-top feeder (or other means for feeding the bees syrup). Keep the entrance wide open (no entrance reducer).

3. **Place a bedsheet in front of the new hive, from the ground to the hive entrance.**

   This ramp will help the bees find the entrance to their new home. In lieu of a bedsheet, you can use a wooden plank or any configuration that creates a gang plank for the bees.

4. **Take the box containing the swarm and shake/pour the bees onto the bedsheet, as close to the entrance as possible.**

   Some of the bees will immediately begin fanning an orientation scent at the entrance, and the rest will scramble right into the hive. What a remarkable sight this is — thousands of bees marching into their new home. Congratulations! You have a new colony of bees!

The swarm of bees (now in their new home) will draw comb quickly because they arrive loaded with honey. Feed them syrup using the hive-top feeder to stimulate wax production. Feeding may not be necessary if the nectar flows are heavy.

**TIP**

In a week, check the hive and see how the bees are doing. See any eggs? If you do, you know the queen is already at work. How many frames of foundation have been drawn into comb? The more the merrier! Is it time to add a second deep (see the "The 70 percent rule" sidebar earlier in this chapter)?

Finding a swarm and starting a new colony are typically more desirable earlier in the season than later. That's because late swarms don't have much time to grow and prosper before the winter sets in. There's an old poem of unknown origin that is well known to beekeepers:

A swarm in May — is worth a load of hay.

A swarm in June — is worth a silver spoon.

A swarm in July — isn't worth a fly.

# Absconding

Absconding is a cruel blow when it happens. One day, you go to the hive and find no one home. Every last bee (or nearly every bee) has packed up and left town. What a horror! Here are some of the typical causes of absconding:

>> **Colony collapse disorder (CCD):** This relatively new phenomenon has devastated honey bee colonies around the world. One day the bees are gone with no evidence as to why. The causes are not yet known for certain, but the problem is being vigorously studied. For more information on CCD, see Chapter 11.

>> **Lack of food:** Make sure your hive has an ample supply of honey. Feed your bees sugar syrup when their honey stores are dangerously low (less than two frames of capped honey) and during serious dearth of nectar.

>> **Loss of queen:** This situation eventually results in a hive with no brood. Always look for evidence of a queen when you inspect your bees. Look for eggs!

>> **Uncomfortable living conditions:** Make sure the hive is situated where it doesn't get too hot or too wet. Overheated or overly wet hives make life unbearable for the colony. Provide ample ventilation and tip the hive forward for good drainage.

>> **Itty-bitty (or not so itty-bitty) pests:** Some hives (particularly weak ones) can become overrun with other insects, such as ants or hive beetles. Even persistent raids from wildlife (skunks, raccoons, and bears, for example) can make life miserable for the bees. See Chapter 13 for tips on dealing with these annoyances.

>> **Mites and disease:** Colonies that are infested with mites or have succumbed to disease may give up and leave town. Take steps to prevent such problems and medicate your bees when the situation demands (see Chapters 12 and 13).

# Where Did the Queen Go?

It's every beekeeper's nightmare: The queen is dead, or gone, or lost. Whatever the reason, if the colony doesn't have a queen, it's doomed. That's why you must confirm that the queen is alive and well at every inspection. If you come to the dismal conclusion that your colony is queenless, you can do two things: Let the colony raise its own queen or introduce a new queen into the colony.

TIP

If you are a new beekeeper, don't panic if you can't find your queen. I get more calls about "lost queens" than any other topic. But most of the time the queen is in the hive. It's just that the new beekeeper has not yet become adept at spotting her. If you can't find the queen (she's easy to miss), look for eggs or very young larvae. That's a sure indication she was there in the last three to five days. That's close enough — beekeeping isn't as precise as rocket science.

## Letting nature take its course

To let the colony create a new queen, it must have occupied queen cells or worker cells with eggs or young larvae. If eggs or young larvae are available, the worker bees will take some of them and start the remarkable process of raising a new queen. When the new virgin queen hatches, she will take her nuptial flight, mate with drones, and return to the hive to begin laying eggs. If no eggs or young larvae are available for the colony to raise a new queen, you must take matters into your own hands and order a new queen from your beekeeping supplier (see next section). Or you can find out how to raise your own queen (see Chapter 14).

REMEMBER

The colony must have eggs or young larvae (a size where the C-shaped larvae fill no more than half of the bottom of the cell) to create its own queen. Older larvae (those that have reached a size to fill the cell) or capped brood are too late into the developmental stage to be transformed into new queens.

TIP

Replacing your queen naturally is certainly interesting, but consider the logistics. The entire process (from egg to laying queen) can take a month. That's a precious amount of time during honey collection season, particularly if you live in a climate with a short summer season. In the interest of productivity, it may be better to take matters into your own hands and order a replacement queen.

## Ordering a replacement queen

A faster solution than the *au naturel* method is to order a replacement queen from your bee supplier. Within a few days, a new, potentially vigorous queen will arrive at your doorstep. She's already mated and ready to start producing brood.

The advantages of ordering a queen are clear:

>> It provides a fast solution to the problem of having a queenless colony.

>> The queen is certain to be fertile.

>> It guarantees the pedigree of your stock. (Queens left to mate in the wild can produce bees with undesirable characteristics, such as a bad temper.)

## Introducing a new queen to the hive

After your queen arrives by mail, you must introduce her into the colony. Doing so can be a little tricky. You can't just pop her in: She's a stranger to the colony, and the bees are sure to kill her. You have to introduce her slowly. The colony needs time to accept her and become accustomed to her scent. Old-time beekeepers swear by all kinds of methods — and some are downright weird. (I don't want you to try them so I'm not going to mention them here!) I suggest that you use one of the following tried-and-true approaches:

1. **Remove one of the frames from the brood box.**

   Pick a frame with little or no brood on it because whatever brood is on the frame will be lost — you won't use this frame again for a week.

2. **Shake all the bees off the frame and put it aside for the next week.**

3. **With the one frame removed, create a space in the center of the brood box. Use this space to hang the queen cage in the same way you hung it when you first installed your package bees (see Chapter 6 and Figure 10-4).**

**FIGURE 10-4:** Hanging a queen cage.

*Courtesy of Howland Blackiston*

**TIP**

Make sure to remove the cork from the queen cage to expose the candy plug. Also, when you hang the cage, make sure the candy end is facing *up*. That way, any attendant bees that die in the cage will not block the hole and prevent the queen from getting out. Leave the bees alone for one week, and then inspect the hive to determine that the queen has been released and that she is laying eggs.

If the weather is mild (over 60 degrees Fahrenheit [16 degrees Celsius] at night), you can introduce the queen cage on the bottom board (see Figure 10-5). Remove the cork to expose the candy plug. Slide the cage screen side up along the bottom board and situate it toward the rear of the hive. Use your hive tool to nudge it as far to the rear of the hive as possible. Leave the bees alone for one week and then inspect the hive to determine that the queen has been released and that she is laying.

**TIP**

Use a flashlight at the entrance to peer into the back of the hive to see if the hole in the cage is clear. If yes, the queen has likely been released.

**FIGURE 10-5:**
Sliding a queen
cage onto the
bottom board.

*Courtesy of Howland Blackiston*

# Avoiding Chilled Brood

Honey bees keep their hive clean and sterile. If a bee dies, the others remove it immediately. If a larva or pupa dies, out it goes. During the early spring, the weather can be unstable in some regions of the country. A cold weather snap can chill and kill some of the developing brood. When this happens, the bees dutifully remove the little corpses and drag them out of the hive. Sometimes the landing board at the entrance is as far as they can carry them. You may spot several dead brood at the entrance or on the ground in front of the hive. Don't be alarmed — the bees are doing their job. A few casualties during the early spring are normal.

**TIP**

Chilled brood looks similar to, but is different from, the disease *chalkbrood*. You can find information about chalkbrood in Chapter 12.

Sometimes beekeepers unwittingly contribute to the problem of chilled brood. Remember, chilled brood is killed brood. You can do a few things to avoid endangering your bees:

>> When the temperature drops below 50 degrees (10 degrees Celsius), keep your inspections very, very brief. A lot of heat escapes every time you open the hive, and brood can quickly become chilled and die.

>> Provide adequate ventilation to avoid condensation, especially at the top and sides of the hive. The resulting icy water dripping on the comb can chill the brood.

>> Inspect your bees only on days when there is little or no wind (especially during cool weather). Harsh winds will chill (and kill) brood.

Extreme close-up of a worker bee's head. Note the bee's tongue is extended.

Beekeeper David Wright holds a branch containing a nice swarm that's ready to put into a new hive. Honey bees are remarkably gentle during swarming

This uber-close image shows in detail the anatomy of a worker (left) and a drone (right). Note the drone's huge eyes and larger body.

Confirm American foulbrood by inserting a toothpick or wooden match into a capped cell. If you note a ropy, gooey mass as you slowly withdraw the stick, you have good reason to believe your colony has AFB.

The spotty brood pattern and the sunken, perforated cappings are reasons to suspect American foulbrood.

2013 Rob Snyder

This brown, shriveled larvae at the bottom of its cell is the result of European foulbrood.

Varroa mites are visible  to the naked eye. Note the  Varroa on this worker bee.

These hard, chalky, mummified pupae indicate chalkbrood disease in this hive.

When the larvae of the small hive beetle hatch, they tunnel through combs, feeding on pollen, honey, and brood and defecating, causing fermentation and leading to what's called "a slime out," as you can see from this slime image.

This remarkable close-up photo illustrates the visual differences between a worker (top), queen (middle), and drone (bottom).

Wax moths can destroy honeycomb in a weak hive, leaving a hopeless, webby mess for the beekeeper. This frame is a goner.

Honey bees gather pollen and store it in pollen baskets located on their hind legs. Pollen is used by bees as a protein food.

Honey's not the only food you can get from your bees. Try your hand at brewing mead (honey wine). Delicious!

These two workers share nectar. This form of socialization helps bees communicate the type of food sources available for foraging.

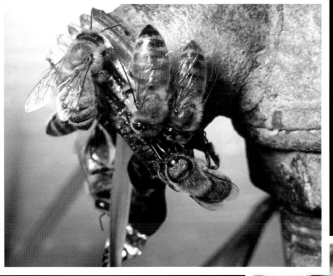

Foraging worker bees gather water from a leaky water spigot. The bees use water to cool the hive and dilute honey.

That's pollen in these cells. The color of the pollen depends upon the flowers from which it originates. In time, you will be able to identify the plants your bees are visiting by color of the pollen they deposit in the cells.

You will soon discover that honey bees are gentle creatures. This beekeeper is hand feeding a drop of sugar syrup to one of his girls.

After grafting larvae into the queen cups, the bees build queen cells; you'll soon have a whole lot of queens underway. Be sure to add a protector around each cell before the queens start emerging.

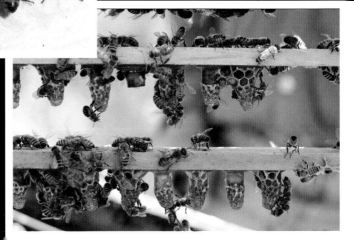

Beekeeping is a great hobby for the entire family. By all means, get the little ones involved. All that energy will come in handy come bottling season.

You can turn queen rearing into a nice little business. Local beekeepers will be eager to find a local source for strong, healthy, productive queens. These queens are caged and ready for sale.

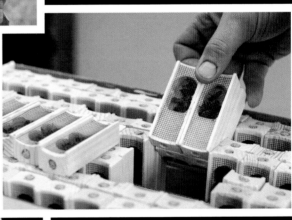

A beautiful frame of capped brood. Note the tight brood pattern with capped honey in the upper third of the frame.

These young worker bees are just emerging from their cells. Note the soft, downy fur on these youngsters.

These two peanut-shaped cells contain developing queens.

You can really see the huge wrap-around eyes on this young drone that is emerging from his cell.

# Dealing with the Dreaded Robbing Frenzies

*Robbing* is a situation in which a hive is attacked by invaders from other hives. The situation is serious for a number of reasons:

>> A hive defending itself against robbing will fight to the death. This battle can result in the loss of many little lives and even destroy an entire colony. Tragedy!

>> If the hive is unable to defend itself in a robbing situation, the invading army can strip the colony of all its food. Disaster!

>> Being robbed changes the disposition of a hive. The bees can become nasty, aggressive, and difficult to deal with. Ouch!

Many new beekeepers mistake a robbing situation as being the opposite of a problem. Look at all that activity around the hive! Business must be booming! It's a natural mistake. The hive's entrance is furious with activity. Bees are everywhere. Thousands of them are darting in, out, and all around the hive. But look more closely. . . .

## Knowing the difference between normal and abnormal (robbing) behavior

A busy hive during the nectar flow may have a lot of activity at the entrance, but the normal behavior of foraging bees looks different than a robbing situation. Foraging bees go to and fro with a purpose. They shoot straight out of the hive and are quickly up and away. Returning foragers are weighted down with nectar and pollen and land solidly when returning to their hive. Some even undershoot the entrance and crash-land just short of the bottom board.

Normal activity at the hive's entrance can look unusually busy. This is when young worker bees take their orientation flights. Facing the hive, they hover up, down, and back and forth. They're orienting themselves to the location of their hive. You may see hundreds of these young bees floating around the front of the hive, but there's nothing aggressive or frantic about their exploratory behavior.

You may also see a lot of activity during the afternoons when the drones are in flight, searching enthusiastically for virgin queens from other colonies. These drone flights are normal frenzies and should not be mistaken for robbing behavior.

In contrast to these normal busy situations, robbing takes on an aggressive and sinister look. Try to recognize the warning signs:

>> Robbing bees approach the hive without being weighted down with nectar. They may not shoot right into the entrance. Instead, they fly from side to side, waiting for an opportune moment to sneak past the guard bees.

>> If you look closely, you may see bees fighting at the entrance or on the ground in front of the hive. They are embraced in mortal combat. These are the guard bees defending their colony to the death. This behavior is a sure indication of robbing.

>> Unlike foraging bees that leave the hive empty-handed, robbing bees leave the hive heavily laden with honey, which makes flying difficult. Robbing bees tend to climb up the front of the hive before taking off. Once they're airborne, there's a characteristic dip in their flight path.

## Putting a stop to a robbing attack

If you think you have a robbing situation under way, don't waste time. Use one or more of the following suggestions to halt robbing and prevent disaster:

>> Reduce the size of the entrance to the width of a single bee. Use your entrance reducer or clumps of grass stuffed along the entrance. Minimizing the entrance will make it far easier for your bees to defend the colony. But be careful. If the temperature has turned hot, narrowing the entrance impairs ventilation.

>> Soak a bedsheet in water and cover the hive that's under attack. The sheet (heavy with water) drapes to the ground and prevents robbing bees from getting to the entrance. The bees in the hive seem to be able to find their way in and out. During hot, dry weather, rewet the sheet as needed. Be sure to remove the sheet after one or two days. By that time the robbing behavior should have stopped.

## Preventing robbing in the first place

The best of all worlds is to prevent robbing from happening at all. Here's what you can do:

>> Never leave honey out in the open where the bees can find it — particularly near the hive and during a dearth in the nectar flow. Easy pickings can set off a robbing situation.

>> When harvesting honey, keep your supers covered after you remove them from the colony.

>> Be very careful when handling sugar syrup. Try not to spill a single drop when feeding your bees. The slightest amount anywhere but in the feeder can trigger disaster.

>> Until your hive is strong enough to defend itself, use the entrance reducer to restrict the size of the opening the bees must protect. Also, be sure to close off the ventilation groove in the inner cover, if yours has one.

>> Never feed your bees in the wide open (such as filling a dish with syrup or honey and putting it near the entrance of the hive).

>> Avoid using a Boardman entrance feeder (see Figure 10-6). Being so close to the entrance, these feeders can sometimes incite robbing behavior.

**FIGURE 10-6:**
I don't advocate using a Board-man entrance feeder — the smell of the syrup can entice strange bees to rob your hive.

*Courtesy of Howland Blackiston*

**WARNING**

Don't be tempted to make it easier for your bees to access the syrup you feed them. I know of a beekeeper who put shims between the hive-top feeder and hive to create a gap that makes it easier for the bees to access the syrup. The result was a furious robbing attack from other bees. Keep your feeding device where only your colony can reach it.

# Ridding Your Hive of the Laying Worker Phenomenon

If your colony loses its queen and is unable to raise a new queen, a strange situation can arise. Without the "queen substance" wafting its way through the hive, there is no pheromone to inhibit the development of the worker bees' reproductive organs. In time, young workers' ovaries begin to produce eggs. But these eggs are not fertile (the workers are incapable of mating). So the eggs can only hatch into drones. You may notice eggs, larvae, and brood and never suspect a problem. But you have a huge problem! In time, the colony will die off without a steady production of new worker bees to gather food and tend to the young. A colony of drones is doomed.

## How to know if you have laying workers

Be on the lookout for a potential laying-workers situation and take action when it happens. The following are key indicators:

>> **You have no queen.** Remember that every inspection starts with a check for a healthy, laying queen. If you have lost your queen, you must replace her. Quickly.

>> **You see lots and lots of drones.** A normal hive never has more than a few hundred drone bees. If you notice a big jump in the drone population, you may have a problem.

   If you look really closely you might notice runty drones being reared in worker cells.

>> **You see cells with two or more eggs.** This is the definitive sign. A queen bee will place only one egg in a cell — seldom more than one. Laying workers are not so particular; they will place two or more eggs in a single cell. Some of these eggs may be placed haphazardly on the side of the cell or only partway into the cell. If you see many cells with more than one egg in them, and some are improperly positioned on the bottom (see Figure 10-7), you can be certain that you have laying worker bees. Time to take action!

## Getting rid of laying workers

You may think that introducing a young and productive queen will set things right. But in a laying worker situation, it won't. The laying workers will not accept a queen once they have started laying eggs. If you attempt to introduce a queen, she will be swiftly killed. Guaranteed.

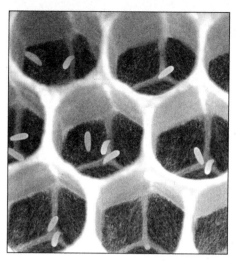

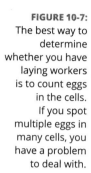

**FIGURE 10-7:**
The best way to determine whether you have laying workers is to count eggs in the cells. If you spot multiple eggs in many cells, you have a problem to deal with.

*Courtesy of Howland Blackiston*

Before you can introduce a new queen, you need to get rid of all the laying workers. But how? They look just like all the other workers! The solution is tedious and time-consuming but 100 percent effective when done properly. You need the following items:

>> An empty deep hive body (no frames). The empty hive body will be used to temporarily hold the frames you remove from the problem hive. You will need two empty hive bodies if your problem hive consists of two deep hive bodies.

>> An outer cover

>> A wheelbarrow or hand truck

Follow these steps:

1. **Order a new marked queen from your bee supplier.**

2. **The day your queen arrives, put the entire "problem" hive (bees and all, minus the bottom board) in the wheelbarrow (or on the hand truck) and move it at least 100 yards away from its original location. You'll want those spare empty hive bodies and outer covers nearby.**

   The bottom board stays in its original location.

3. **One by one, shake every last bee off each frame and onto the grass.**

   Not a single bee can remain on the frame — that bee might be a laying worker. A bee brush (see Chapter 5) helps get the stubborn ones off the frames.

4. **Put each empty frame (without bees) into the spare empty hive(s) you have standing by. These should be at least 15 to 20 feet away from the shaking point.**

   Make sure that no bees return to these empty frames while you are doing the procedure. Use the extra outer cover to ensure that they can't sneak back to their denuded frames.

5. **When you have removed every bee from every frame, use the wheelbarrow or hand truck to return the old (now bee-less) frames to the original hive bodies.**

   Again, make sure that no bees sneak back onto the frames.

6. **Place the hive in its original location on the bottom board, and transfer all the beeless frames from their temporary housing. So now you have the original hive bodies back at their original location, and all the original frames (less any bees) placed back into the hive.**

   Some of the bees will be there waiting for you. These are the older foraging bees (not the younger laying workers). Be careful not to squash any bees as you slide the hive back onto the bottom board.

Most of these older foraging bees will find their way back to the hive. But the young nurse bees, the ones that have been laying eggs, have never ventured out of the hive before. They will be lost in the grass where you deposited them and will never find their way back to the hive. They are goners.

Now you can safely introduce your new queen using a queen cage. See the instructions earlier in this chapter on how to introduce a new queen.

There are rare occasions when a queen will lay nothing but drone eggs. This happens when a new queen does not successfully mate. Queens that don't mate (or older queens that use up their stored sperm) can only produce unfertilized (drone) eggs. Such queens are termed drone layers. Such colonies are doomed because foraging workers are needed to survive and fertilized eggs are necessary to raise a replacement queen.

# Preventing Pesticide Poisoning

With what we are finding out about colony collapse disorder and its potential relationship to pesticides, we can't be too careful when it comes to pesticide use. I get upset when I see people spraying their lawns and trees with pesticides. These chemicals may make for showcase lawns and specimen foliage, but they are no good for the water table, birds, earthworms, and other critters. Some of these treatments are deadly to bees. (*Note:* I'm not talking about fertilizers, just pesticides here.) If you ever see a huge pile of dead bees in front of your hive, you can be pretty sure that your girls were the victims of pesticide poisoning. Here are a few things you can do to avoid such a tragedy:

» Let your neighbors know that you are keeping bees. Make sure they know how beneficial pollinating bees are to the community and ecology. Explain to them the devastating effect that pesticide spraying can have on a colony. They may think twice about doing it at all. If they must spray, urge them to do so at dawn or dusk, when the bees are not foraging. Encourage your neighbors to call you the day before they plan to spray. With advance warning, you can protect your bees.

» On the day your neighbors plan to spray, place a towel that has been saturated with water on top of the outer cover. This will be a water source for the colony. Then cover the entire hive with a bedsheet that you have saturated with water to give it some weight. Let it drape to the ground. The sheet will minimize the number of bees that fly that day. Remove the sheet and towel the following morning after the danger has passed.

**TIP**

Alternatively, you can screen the entrance the night before the spraying and keep the girls at home the entire day. Remove the screen and let them fly the next day.

» Register your colony with your state's department of agriculture. You may have to pay a minimal charge for registration. Some states publish a list of all registered beekeepers in the state. Reputable arborists check such lists before spraying in a community. If you are on the list, they will hopefully call you before they spray in your area.

# The "Killer Bee" Phenomenon

The media has had a ball with the so-called "killer bees." These nasty-tempered bees have been fodder for fantastic headlines and low-budget horror movies. At the same time, this kind of publicity has had a negative and unwelcome impact on backyard beekeeping. The resulting fear in the community can make it difficult for a beekeeper to gain the support and acceptance of his neighbors. Moreover, sensational headlines have resulted in sensational legislation against keeping bees in some communities. The public has been put on guard.

"Killer bees" present another problem for the beekeeper as well: If your area has them, you must manage your colony extra carefully to prevent your own bees from hybridizing and becoming more aggressive.

# What are "killer bees"?

First of all, let's get the name correct. The bees with the bad PR are actually Africanized honey bees (AHB) — or *Apis mellifera scutellata* if you want to get technical. The "killer bee" pseudonym was the doing of our friends in the media.

How did the AHB problem come about? It all started in 1956 in Brazil. A well-respected geneticist was experimenting with breeding a new hybrid that he hoped would result in superior honey production. He bred the notoriously defensive honey bee from Africa with the far more docile European honey bee. But a little accident happened. Some African queen bees escaped into the jungles of Brazil. The testy queens interbred with bees in the area, and voilà — the AHB become a force to deal with.

Outwardly, AHBs look just like our friendly European honey bees. In fact, you must take a peek under the microscope or do a DNA test to detect the difference. Their venom is no more powerful. And like our sweet bees, they too die after inflicting a sting. The main and most infamous difference is their temperament. They are very defensive of their hives and are quick to become disturbed. Once alerted, individual bees may chase an intruder long distances, and stay hyper-defensive for days after an incident.

There have been reports of human deaths resulting from attacks by AHBs. But these reports are rare and frequently involve elderly victims who have been unable to fend off the attackers or make a fast getaway. The media can put quite a sensational spin on such tragedies, and that has contributed to some bad PR for honey bees in general.

My friend Kate Solomon, shown in Figure 10-8, worked for several years in the Peace Corps teaching South American beekeepers how to work with the AHB. Kate's efforts (and yes, she puts cotton in her nose and ears to keep unwanted explorers at bay!) resulted in not a single sting from these "killer bees."

# AN EXPERIMENT THAT FLOPPED

Despite the hopes of those Brazilian scientists over half a century ago, the AHB has turned out to be kind of a dud as compared to her European cousins:

- It is more difficult for beekeepers to manage.

- It defends in larger groups, inflicting many more stings.

- It swarms far more frequently.

- It doesn't produce appreciably more honey.

- It makes less wax.

**FIGURE 10-8:**
This young lady allows herself to be covered by so-called "killer bees." She did not receive a single sting during this *what-in-the-world-was-she-thinking* demonstration. My advice? Don't try this at home!

*Courtesy of Kate Solomon*

## Bee prepared!

In the half-century since "the accident" in Brazil, AHBs have been making their way northward to the United States. In 1990, the first colonies of AHBs were identified in southern Texas. As of this writing, they have been verified through all the southern states. There is speculation as to how far north these bees are capable of surviving (after all, they are a tropical species). In any event, they have arrived amid great publicity. Beekeepers and the public will have to learn how to deal with them. For an up-to-date map of the progress, go online to `www.usda.gov` and search "Africanized Bees."

TIP

Here are some helpful hints about safe beekeeping in areas known to be populated by AHBs:

>> If you live in an area where AHBs have been seen, do not capture swarms or populate your hive with anything other than package bees from a reputable supplier. Otherwise, you may wind up with the hive from hell.

>> If you are unlucky enough to disturb a colony of AHBs, don't stick around to see how many will sting you. Run in a straight line far away from the bees. AHB are fast flyers, and you will have your work cut out for you when you attempt to outrun them. Don't jump into water — they'll be waiting for you when you surface. Instead, enter a building and stay inside until things cool off.

>> In the areas where the AHB has been introduced, diligent beekeepers are the community's best defense against the AHB's spread. By systematically inspecting her hive to spot her *marked* queen, a beekeeper knows that her colony remains pure. Only when an unfamiliar queen (perhaps an AHB) is introduced is the colony's genetic integrity at risk. More than ever, backyard beekeepers are needed to ensure that the AHB doesn't become a problem in any community.

>> If you join a local bee club (and I highly recommend that you do), encourage the club to publish information educating the public about the benefits of beekeeping. Teach the community the real story about the AHB. Take positive steps to quell the fear that may lurk in some people's minds. Let them know how important it is to have beekeepers who can help control the spread of the AHB. A good education program is a beekeeper's best defense against local legislation restricting beekeeping in the community. And please don't call them "killer bees."

Chapter **11**

# Colony Collapse Disorder

*S*pecial thanks to the USDA and Dr. Dewey Caron for their help with this chapter.

Unless you have been living in a remote cabin on the side of a forgotten hill, you have likely noted that the media has been abuzz with news about "the vanishing bees." The last few years have been unhappy ones for our bees. Since 2006, thousands upon thousands of honey bee colonies have been vanishing. Gone without a trace. Poof!

Colony collapse disorder (CCD) is the name that was initially given to what seems to be the most serious die-off of honey bee colonies in decades. And to get right to the point, after years of study, it's not known with precision what is causing it.

Although this news is unhappy, don't despair. Becoming a new beekeeper is one of the most useful things you can do to help save our lovely honey bees. More on what else you can do later in this chapter.

# What Is CCD?

In the autumn of 2006, a beekeeper in Florida filed the first report of a sudden and unexplained disappearance of his bees. They didn't die. They just packed up and left. More reports of heavy losses (mostly from commercial migratory beekeepers) quickly followed. In subsequent years, beekeepers have reported losing anywhere from 30 percent to 90 percent of their hives. Like a firestorm, this tragedy has swept across nearly all of the United States as well as some countries overseas. It has affected both commercial beekeepers and hobbyists. It is a far-reaching problem that has serious consequences (see the upcoming section "Why All the Fuss?").

Colony collapse disorder (CCD) is characterized by the sudden and unexplained disappearance of all adult honey bees in the hive, usually in the fall. In one scenario, a few young bees and perhaps the queen may remain behind while the adults disappear. Or in another scenario, there may be no bees left in the hive. Honey and pollen are usually present, and there is often evidence of recent brood rearing. This abrupt evacuation is ordinarily highly unusual because bees are not inclined to leave a hive if there is brood present.

Another puzzling characteristic is that opportunists (such as robbing bees from other hives, wax moths, and small hive beetles) are slow to invade colonies experiencing CCD. There are no adult bees present to guard the hive and lots of goodies to loot, yet these invaders stay clear. Hmmm. What do they know that we don't?

Sometimes (rarely) bees abscond from a hive because conditions are too unpleasant to remain in the hive: too hot, too many pests, not enough food, no queen, and so on (see Chapter 10). But CCD is different from such absconding. Conditions don't appear to be unfavorable. And it's happening at an alarming rate.

Colonies that experience CCD have the following characteristics:

>> All or nearly all of the bees pack up and leave within a two- to four-week time period. But there are no dead adult bodies.

>> In some instances the queen and a small number of young-aged survivor bees are present in the brood nest. There are no or very few dead bees in the hive or at the hive entrances.

>> Capped brood is left behind.

>> There is stored pollen and capped honey.

>> Empty hives are *not* quickly invaded by opportunists (robbing bees, wax moths, small hive beetles, and so on).

# What to Do If You Suspect CCD

If you think your hives have fallen to CCD, don't panic. As a new beekeeper, you may be jumping to unwarranted conclusions. To date, CCD has been far more prevalent among commercial beekeepers, although losses among hobbyists like us happen sometimes.

**TIP**

If you believe you may have a problem, I urge you to call your state's department of agriculture and ask to speak with the head bee inspector. He or she is likely to provide you with some helpful information. If you have records of the number of mites in your hive prior to the collapse, it will be helpful for the inspector to know (see Chapter 13 for more about mites).

# Why All the Fuss?

The media is all over this story. It has hit the evening national news. It's made the cover of *Time* magazine. Why is CCD making headlines? Imagine a world without bees. That would be an unhappy world. Did you know that honey bees account for 30 percent of everything you eat? Commercial beekeepers provide honey bees to farmers all around the country to pollinate the crops that wind up in our super-markets. If these pollinating mavericks were all to disappear, there would be reduced crop variety and most likely higher prices in your grocery store. No question about it. Honey bees are critical for agricultural pollination — adding more than $15 billion in value to about 130 crops — especially crops like berries, nuts, fruits, and vegetables. The unexplained disappearance of so many colonies is not a matter to take lightly. Table 11-1 summarizes some of the consequences of a world without bees.

**TABLE 11-1**     **Pollination Experiments**

| Crop | Without Honey Bees | With Honey Bees |
|---|---|---|
| Pears | 99 lbs. fruit | 344 lbs. fruit |
| Alfalfa | 62 lbs. seed per acre | 220 lbs. seed per acre |
| Apples | 25 apples per tree | 1,200 apples per tree |

*Experiment results showing comparison of crops pollinated by honey bees versus the same crops netted to prevent pollination.* **Source:** *W.R. Roach Company Orchards and other sources.*

# What's Causing CCD?

No one knows for certain what's causing CCD, at least not at the time of this writing. But researchers have managed to dismiss some "wild" theories and are now focusing on other, more probable causes. In all likelihood, CCD is not due to a single factor.

## The cellphone theory

There was a well-publicized theory that the explosion of cellphone usage was causing CCD. Could it be? The short answer is no.

A small study done in Germany seemed to indicate that a particular type of base station for mobile phones could screw up honey bee navigation. But, despite all the media attention that this study received, it had nothing to do with CCD. The researcher who conducted the study told the Associated Press that there is "no link between our tiny little study and the CCD-phenomenon . . . anything else said or written is a lie." The scientists studying CCD agree. Case closed.

## It may be the perfect storm

Far more likely it is not one single thing that is causing CCD, but rather a brew of many different challenges that have contributed to this problem. In a nutshell, several potential causes are being studied by scientists around the world: parasites (such as mites), pathogens (disease), environmental stresses (which include pesticides), and management stresses (including nutrition problems). If CCD is a combination of factors, it makes investigating the root cause especially complex. There are so many variables!

Although this is not a complete list of what's being studied, here are some of the more significant ingredients to this dire cocktail.

### Parasites

The spread of Varroa and tracheal mites has seriously affected honey-bee health in the United States and around the world. Varroa had nearly wiped out honey bees in the wild (feral hives), although these populations are now recovering. Both mites have put a major stress on our honey bees and could certainly make our girls far more susceptible to some of the other causes being studied.

### Pathogens

In Chapter 12, I talk about bee viruses. The Varroa mite has been shown to spread several different viruses among honey bees as it feeds on adult and pupal bees.

Although many different viruses can impact honey-bee health, a few in particular are being studied in connection with CCD.

One is *Israeli acute paralysis virus* (IAPV). This particular virus is not necessarily the cause of CCD, but is more likely one ingredient to that cocktail that might trigger CCD. This research is being headed by Dr. Diana Cox-Foster at Penn State College of Agricultural Sciences. For the most up-to-date information on the potential connection between CCD and IAPV, visit http://cas.psu.edu/.

Deformed wing virus seems to have increased in its prevalence and is one that appears particularly deadly to honey bees.

Other viruses, such as the acute bee paralysis virus, chronic paralysis virus, Kashmir bee virus, black queen cell virus, and sacbrood virus, also contribute to some degree and cause honey-bee viral disease epidemics in different colonies.

Another virus that recently gained attention is called *tobacco ringspot virus* (TRSV). Researchers have found this virus (that typically infects plants) has been systemically infecting honey bees. The rapidly mutating virus jumped from tobacco plants to soy plants to bees. The study provides the first evidence that honey bees exposed to virus-contaminated pollen can also be infected and that the infection becomes widespread in their bodies. Researchers state that honey bees can transmit TRSV when they move from flower to flower during the pollination process. The virus may be causing systemic infection in honey bees.

I know, I know. This is all sounding gloomy and overwhelming. But rest assured that smart people are working diligently on remedies. Researchers are exploring the ways in which these viral cocktails not only travel and migrate, but how they systemically infect the bodies of honey bees and lead to the eventual collapse of hives.

## Pesticides

Another factor in bee loss is pesticides. Researchers have found higher-than-expected levels of miticides (used and sometimes misused by beekeepers to control mites) plus traces of a wide variety of agricultural chemicals in the pollen and wax of inspected hives.

Some believe that pesticides, especially a relatively new class called *neonicotinoids*, may have a role in CCD. Neonicotinoids are known to be toxic to bees. They can impair olfaction memory, motor activity, feeding behavior, and the bees' navigation and orientation. Neonicotinoids and some fungicides are synergistic, meaning that after exposure to one type of chemical, subsequent exposure to another chemical results in a far more toxic situation.

What is suspected — and probably the most damaging — is the sublethal or chronic effect of neonicotinoids. Exposure to continued levels of pesticides over time may render the bees weakened and thus more susceptible to viral infections that can then decimate the colony. It is like two body blows; the bees just can't take the one-two punch. All in all, this is some nasty stuff. Pesticide involvement in CCD remains a strong possibility.

Most pesticides that are acutely toxic to bees kill the adults, first foragers, and sometimes the hive bees. The foragers die in the field. With some chemical pesticides, large numbers of dead bodies may be evident at the front of the hive (see Figure 11-1). A small number of pesticides are even more damaging when they are brought back to the hive in contaminated pollen, causing death of the nurse-aged bees and the brood, which are then not adequately tended to because of fewer nurse bees. Adult bodies and brood pile up quickly and in large numbers in front of the entrance or on the bottom board. Colonies may lose a whole generation of brood rearing, setting back the development of the expanding colony.

CCD, on the other hand, is different. No dead bodies are found in or around the hive. The bees die away from their hive.

**FIGURE 11-1:** The huge pile of dead bees in front of this hive is a telltale indication of pesticide poisoning.

*Courtesy of Katie Lee, Bee Informed Partnership*

## Other possibilities

A host of other possible causes is now under study, including the following:

>> Nutritional fitness of the adult bees

>> Level of stress in adult bees as indicated by stress-induced proteins

>> The use of honey-bee antibiotics (especially new products in the market)

>> Feeding bees high-fructose corn syrup (as is common with commercial beekeepers)

>> Availability and quality of natural food sources

>> Lack of genetic diversity and lineage of bees

The nutritional factor, resulting in stress on bees, seems to be a factor in weak colonies or colonies not developing properly. This can be because of loss of weeds in agricultural fields (because of increasingly heavy use of herbicides), monocultures of only single flowering sources for bee nutrition (as when the bees are involved in commercial pollination), effects on the bees' intestinal microflora (because of heavy reliance on artificial feeding of colonies), and decreasing natural forage (because of human disturbance of the habitat where bees need to forage). Bees that are weakened because of one or more of these factors may become vulnerable to the other factors.

All in all, honey-bee colonies are suffering from a toxic whammy of multiple negatives.

# Answers to FAQs

**Is honey from CCD colonies safe to eat?** To date, there is no evidence that CCD affects honey. The impact of CCD appears to be limited to adult bees.

**Is it safe to reuse the equipment from colonies that are lost during the winter?** If it can be determined that the bees starved or died because of other reasons associated with typical winter loss (such as mites), it is safe to reuse equipment, including the remaining honey and pollen. In addition, if your colonies died from what appears to be CCD, reusing equipment is still okay. Just allow your equipment to air out for a few days before reusing.

**Who is working on this problem?** An army of researchers, apiculture extension specialists, and government officials have come together to work on CCD. This particular group is called the Mid Atlantic Apiculture Research Extension Consortium (MAAREC). For up-to-date information on the research, visit the website: http://agdev.anr.udel.edu/maarec/.

The Honey Bee Health Coalition is a broad-based, diverse assembly of beekeeping organizations, commodity and specialty crop producers, agro-business, supply chain companies, non-government organizations, universities, and agencies that promote a vision of "Healthy Bees, Healthy People, and Healthy Planet." Consult its website at www.honeybeehealthcoalition.org.

Additional information on CCD can be found at the website for the United States Department of Agriculture, www.usda.gov/wps/portal/usdahome.

# What You Can Do to Help

There are a lot of things you can do to fend off CCD in your neck of the woods. Although we don't yet know the actual causes of CCD, here are some sensible actions you can take immediately:

» Become a beekeeper! What a great way to reintroduce honey bees in your area. Hopefully, this book will get you started.

» Keep colonies strong by practicing best-management practices. In other words, follow the steps in this book religiously!

» Replace old brood comb with new foundation every three years. This will minimize the amount of residual chemicals/pathogens that may be present in old wax.

» Avoid unduly stressing your colonies. Seek to provide adequate ventilation; feed your bees when pollen and nectar are scarce; keep mite infestations in check; and medicate against Nosema (see Chapter 12).

» Monitor Varroa mite populations and take steps to treat your colony when mite levels become unacceptable. Visit www.honeybeehealthcoaltion.org/varroa for guidelines for Varroa management. Also, look at Chapter 13 for more information on Varroa mites.

» Always practice an integrated pest management (IPM) approach for Varroa control in honey-bee colonies. This approach can minimize the need for chemical use in your hives and lessen the bees' exposure to chemicals (see the "What is IPM?" sidebar later in this chapter).

>> Avoid the use of chemicals and pesticides in your garden and on your lawn. The use and misuse of pesticides is on the short list of factors that may be harming honey bees. Limit the use of these chemicals, or better yet, go au naturel (after all, my dandelion lawn is beautiful!). Convince your neighbors to do the same.

>> Plant a bee-friendly garden. Good nutrition is vital to the overall health of the colony. See Chapter 18 for some ideas of flowering plants that your bees will love.

>> Write your congressional representatives. Funding for honey-bee research is more critical than ever. Let the feds know you care about our precious honey bees.

Don't let all this gloomy news hinder your enjoyment of beekeeping. Although CCD is a serious concern for our honey bees, I am confident that remedies will be forthcoming. As mentioned earlier, becoming a backyard beekeeper is the single best thing you can do to help our honey bees. Embrace and enjoy this glorious hobby, and feel good about helping the honey bee get back on its feet. All six of them.

**ALL NATURAL**

Scattered throughout this book is information that features "natural" approaches for keeping your bees healthy. Be on the lookout for these best-practice examples that minimize or eliminate the use of chemicals in your colonies.

## WHAT IS IPM?

The idea of IPM (integrated pest management) is to manage honey-bee pests (such as wax moths, Varroa, and tracheal mites) by using a wider variety of control techniques that seek to minimize the use of toxic chemicals. The key word is *manage,* not necessarily eliminate the pests. Some general rules include the following:

- Prevention is better than attempting a cure — just do what's needed to help the bees help themselves.

- Monitor, sample, and test regularly to make sound decisions.

- Use soft (nonchemical) treatments whenever possible.

- Use hard treatments (medications and chemicals) only when absolutely necessary and as a last resort.

- When chemicals must be used, practice rotational use of chemicals to avoid pests developing resistance.

*(continued)*

*(continued)*

A whole book could be devoted to the nuances of IPM, but there are many examples of IPM scattered throughout this book. Consider the following as IPM best-practice techniques:

- Place honey supers in the freezer before storage to kill wax moth larvae and pupa.

- Use screened-bottom boards to help manage Varroa populations.

- Develop resistant bee stock by raising your own queens from your heartiest colonies. Or at the very least, when re-queening, select bee stock with higher hygienic behaviors.

- Use drone comb to capture and remove Varroa mites as a trapping mechanism during spring buildup.

- Use soft chemicals for mite control, such as essential oils, and acids (for Varroa) or menthol and grease patties (for tracheal mites).

- Replace old wax comb with new foundation every two to three years.

- Going into winter, place metal mouse guards at the entrance of the hive.

- Feed colonies sugar syrup to stimulate spring buildup and when foraging resources are in short supply.

- After the nectar flow, consider splitting a large colony into two colonies to help reduce mite numbers in the smaller, developing splits. When doing so, be sure to re-queen with Varroa-resistant stock (see Chapter 18 for information on dividing a colony).

# Chapter **12**

# Understanding Diseases and Remedies

won't pretend otherwise — this is *not* the fun part of beekeeping. I'd much rather never have to think about my bees getting sick. My heart aches when they do. Nothing is more devastating than losing a colony to disease. But let's get real. Honey bees, like any other living creatures, are susceptible to illness. Although some of these diseases aren't too serious, some can be devastating. The good news is that you can prevent many honey-bee health problems *before* they happen, and you can often head off disaster if you know the early signs of trouble.

Right away let me clear up one thing. None of the health problems that affect bees have any impact on human health. These diseases are 100 percent unique to your bees. They're not harmful or contagious in any way to you or your family. Phew! That's a relief!

In this chapter, I highlight the most common health problems that your bees may face. As you inspect your hives, look carefully at the capped and open brood cells (what's going on in these cells is often the barometer of your colony's health). Discover how to recognize the telltale indications of health problems.

**TIP**

This chapter includes information about bee medications. But medications are a subject that doesn't stand still. New products are introduced all the time, and existing products can fall out of favor or be discontinued. On top of that, not all products are legal in all states or all countries. So medications are a moving target, and you should check with your favorite beekeeping supplier for the latest word on what's available and how to use it.

# Medicating or Not?

I know what you're thinking. Should you put medication in your hive or not? Wouldn't keeping everything natural and avoiding the use of any chemicals, medications, or antibiotics be better? Maybe you can even save a few dollars. Or should you just take on the attitude of going for the "survival of the fittest"?

Well, perhaps the answers to these questions depend on your practice in other areas. Do you avoid taking your dog to the vet for distemper shots and heartworm pills? Would you withhold antibiotics if your child came down with bronchitis? Probably not. Bees are no different. Without some help from you, I can assure you they'll eventually have a problem. You may even run the risk of losing your hive entirely. It need not be that way.

Follow a sensible health-check regime and look carefully for signs of trouble every time you inspect your colony. If things get serious, be prepared to judiciously apply remedies to prevent losing your colonies.

TIP

Here's the real key . . . be thoughtful regarding medicating. Unknowingly, many beekeepers have overdone it when it comes to bee health, administering medications "just in case" they might get sick. Don't do it. This kind of prophylactic regime has contributed to an increasing ineffectiveness of meds, antibiotics, and insecticides over recent years. If you choose to medicate, do so only when really necessary.

WARNING

Remember that you should never, ever medicate your bees when you have honey on the hive that is intended for human consumption. If you decide to medicate, do so *before* honey supers go on the hive, or *after* they are removed. For a description of honey supers and their use, see Chapter 5.

# Knowing the Big-Six Bee Diseases

You should be on the lookout for six honey-bee diseases. Others are out there, but these six are the most common you may face. Some are rare, and it's doubtful that you'll ever encounter them. Some are more commonplace (like Nosema and chalkbrood), and knowing what to do if they come knocking is important. One, American foulbrood, is very serious, and you need to know how to recognize and deal with it quickly.

**REMEMBER**

Each time you inspect the brood area of your colony, you're looking for two things: evidence of the queen (look for her, or look for her eggs) and evidence of health problems (look for the symptoms I describe later in this chapter).

# American foulbrood (AFB)

I start with the worst of the lot. American foulbrood (AFB) is a nasty bacterial disease that attacks larvae and pupae. This serious threat is highly contagious to bees (not people) and, left unchecked, is certain to kill your entire colony. It's the most terrible of the bee diseases. Some symptoms are

>> Infected larvae change color from a healthy pearly white to tan or dark brown and die *after* they're capped.

>> Cappings of dead brood sink inward (becoming concave) and often appear perforated with tiny holes.

>> The capped brood pattern no longer is compact, but becomes spotty and random. This is sometimes referred to as a "shotgun" pattern (see this book's color-photo section).

>> The surface of the cappings may appear wet or greasy.

**TIP**

If you see these conditions, confirm that it's AFB by thrusting a toothpick or matchstick into the dead brood, mixing it around, and then *slowly* withdrawing the toothpick. Observe the material that is being drawn out of the cell as you withdraw the toothpick. Brood killed by AFB will be stringy and will *rope out* about ¼ inch (like pulling taffy) and then snap back like a rubber band (see this book's color-photo section). That test can confirm the presence of AFB. Also take a close look at the dead pupae. Some may have tongues protruded at a right angle to the cell wall. There may also be a telltale odor associated with this disease. Most describe it as an unpleasant, foul smell. If you detect a foul smell and that smell lingers in your nose after leaving the hive, your bees might have AFB.

If you suspect that your bees actually *have* AFB, immediately ask your state bee inspector or a seasoned local beekeeper to check your diagnosis. Treatment for AFB is subject to state law in the United States. If AFB is rampant, it is likely that your hives and equipment will have to be burned and destroyed. Why such drastic measures? Sleeping spores of AFB can remain active (even on old, unused equipment) for up to 70 years.

If you are open to the idea of using medications, three products are available (by prescription from a veterinarian): Terramycin (oxytetracycline), Tylan (tylosin tartrate), and Lincomix (lincomycin hydrochloride). To administer, carefully follow the instructions provided with these meds.

**WARNING**

Be wary of purchasing old, used equipment, no matter how tempting the offer may be or no matter how well you know the seller. If the bees that once lived in that hive ever had AFB, the disease-causing spores will remain in the equipment for decades. No amount of scrubbing, washing, sanding, or cleaning can remedy the situation. Please start your new adventure in beekeeping by purchasing new and hygienic equipment.

## European foulbrood (EFB)

European foulbrood (EFB) is a bacterial disease of larvae. Unlike AFB, larvae infected with EFB die *before* they're capped. It's not as horrific as AFB, but it's a problem that should be dealt with if you see signs of it. Symptoms of EFB include the following:

» Very spotty brood pattern (many empty cells scattered among the capped brood). This is sometimes referred to as a "shotgun" pattern.

» Infected larvae are twisted in their cells like an inverted corkscrew. The larvae are either a light tan or brown color and have a smooth, "melted" appearance (see this book's color-photo section). Remember that normal, healthy larvae are a glistening, bright white color.

» With EFB, nearly all the larvae die in their cells before they are capped. This makes it easy for you to see the discolored larvae.

» Capped cells may be sunken in and perforated, but the "toothpick test" won't result in the telltale ropy trail as described previously for AFB.

» A sour odor may be present (but not as foul as that of AFB).

**TIP**

Here's the best way to view frames for diseased larvae. Hold the frame by the ends of the top bar. Stand with your back to the sun and the light shining over your shoulder and down into the cells. The frame should be sharply angled so you are looking at the true *bottom* of the cell. Viewed at this angle, unhealthy brood (and the queen) are often easier to spot. Most new beekeepers interpret the *bottom* as the midrib of the comb. It isn't. The true bottom of the cell is the lower wall of the cell (the wall that's closest to the hive's bottom board when the frame is hanging in the hive).

Because EFB bacteria don't form persistent spores, this disease isn't as dangerous as AFB. Colonies with EFB sometimes recover by themselves after a good nectar flow begins. Although serious, EFB is *not* as devastating as AFB. There's no medical treatment for curing EFB once it has infected a colony, although there are products on the market (such as Tera Bee) that are sold as a preventative of EFB.

**ALL NATURAL**

If you've detected EFB, requeen your colony (replace the old queen with a new one; see Chapter 10) to break the brood cycle and allow the colony time to remove infected larvae. You can even help the bees by removing as many of the infected larvae as you can using a pair of tweezers.

**ALL NATURAL**

It's a good hygienic practice to replace several frames of comb every year. There are compelling reasons for doing this: Replacing old frames/comb minimizes the spread of disease, and old wax can contain residual medication from past treatments — building resistance and making medication treatments ineffective when they are really needed.

# Chalkbrood

Chalkbrood is a common fungal disease that affects bee larvae. Chalkbrood pops up most frequently during damp conditions in early spring. It is rather common and usually not that serious. Infected larvae turn a chalky white color, become hard, and may occasionally turn black. You may not even know that your bees have it until you spot the chalky carcasses on the hive's "front porch." Worker bees on "undertaker duty" attempt to remove the chalkbrood as quickly as possible, often dropping their heavy loads at the entrance or on the ground in front of the hive (see this book's color-photo section).

No medical treatment is necessary for chalkbrood; your colony should recover okay on its own. But you can help your bees out by removing mummified carcasses from the hive's entrance and from the ground around the hive. Also, usually just one frame will have most of the chalkbrood cells.

**TIP**

Misdiagnosing this disease is common because it's easily confused with chilled brood (see Chapter 9). You see carcasses at the hive entrance with both anomalies, but with chalkbrood, the bodies are hard and chalky (not soft and translucent as is the case with chilled brood).

**ALL NATURAL**

Remove the frame containing the majority of chalkbrood mummies from the hive and replace it with a new frame and foundation. This basic sanitation minimizes the bees' job of cleaning up. Also consider replacing your queen by ordering a new one from your bee supplier (or by providing one of your own if you are raising queens — see Chapter 14).

## Sacbrood

Sacbrood is a viral disease of brood similar to a common cold. It isn't considered a serious threat to the colony. Infected larvae turn yellow and eventually dark brown. They're easily removed from their cells, because they appear to be in a water-filled sack. Now you know where the name comes from!

### HONEY-BEE VIRUSES

Adult honey bees may occasionally fall prey to various kinds of viruses. Sacbrood is likely the best known of the bee viruses, but there are quite a few others that impact honey bees (too many to list here). Viruses aren't easily detected and are often overlooked by hobbyist beekeepers. Some researchers are exploring a link between certain viruses and colony collapse disorder (see Chapter 11).

**Note:** Colonies infested with mites (see Chapter 13) are far more susceptible to viral diseases, likely because open wounds created by mites are an invitation to promote viral pathogen growth within the bee's body.

The subject of bee viruses is a fast-moving target, with new developments and discoveries taking place every year. I urge you to subscribe to one or more of the bee journals to keep tabs on developments regarding honey-bee health.

No single medical treatment exists for all honey-bee viruses, but if you know your bees have a virus, you can help. One by one, remove each frame from the hive and carry it 10 to 20 feet away. Now shake all the bees off the frame and return it (empty) to the hive. Do this for all frames. The sick bees will not be able to return to the hive. The healthy ones will have no trouble making it home.

If you have more than one hive, think twice before shaking sick bees onto the ground and exposing other healthy bees to the problem.

No recommended medical treatment exists for sacbrood. But you can shorten the duration of this condition by removing the sacs with a pair of tweezers. Other than that intervention, let the bees slug it out for themselves.

**ALL NATURAL**

Do your best to keep your bees free of stressful problems (mites, poor ventilation, crowded conditions) and they'll have an easier time staying healthy and avoiding diseases such as chalkbrood, sacbrood, and EFB. Be sure to feed them sugar syrup and pollen substitute in the spring and sugar syrup in the autumn.

# Stonebrood

Stonebrood is a fungal disease that affects larvae and pupae. It is rare and doesn't often show up. Stonebrood causes the mummification of brood. Mummies are hard and solid (not sponge-like and chalky as with chalkbrood). Some brood may become covered with a powdery green fungus.

No medical treatment is needed for stonebrood. In most instances worker bees remove dead brood, and the colony recovers on its own.

**ALL NATURAL**

You can help things along by cleaning up mummies at the entrance and around the hive, and removing heavily infested frames (see treatment for chalkbrood). Also, feeding your bees and providing good ventilation to keep moisture under control are good basic sanitary practices.

# Nosema

Nosema is the most widespread of adult honey-bee diseases and contributes increasingly to the weakening or demise of colonies. It was recently determined that there are actually *two* kinds of Nosema, and they are quite different from each other.

## Nosema apis

*Nosema apis* has been around for a long time. This form of Nosema is caused by a small, single-cell parasite, but is classified as a fungus. It's likely the kind of Nosema you will find mentioned in many beekeeping books. Nosema apis affects the intestinal tracks of adult bees — it's kind of like dysentery in humans. It can weaken a hive and reduce honey production by between 40 and 50 percent. It can even wipe out a colony of bees. It's most common in spring after bees have been confined to the hive during the winter.

The problem is that by the time the symptoms are visible, it has gone too far and is difficult or impossible to treat. Some symptoms of Nosema apis are as follows:

- In the spring, infected colonies build up slowly or perhaps not at all.

- Bees appear weak and may shiver and crawl aimlessly around the front of the hive.

- The hive has a characteristic *spotting,* which refers to many streaks of mustard-brown feces that appear in and on the hive. This is an easy-to-identify indication of Nosema apis (although other things can cause a dysentery-like condition).

**ALL NATURAL**

You can discourage Nosema apis by selecting hive sites that have good airflow and a nearby source of fresh, clean water. Avoid damp, cold conditions that can encourage the fungus. Provide your hives with full or dappled sunlight. Creating an upper entrance for the bees during winter improves ventilation and discourages Nosema apis. Purchase your bees and queens from reputable suppliers.

**WARNING**

Some books will tell you that it's possible to medicate for Nosema apis prophylactically (preventively) by feeding Fumigilin-B (an antibiotic) in sugar syrup in the spring and fall. And that's true. But scientists now know that if your bees have the *other* kind of Nosema (Nosema ceranae), medicating with Fumigilin-B may actually exacerbate the ceranae infection rather than suppress it! At least that's what some in-the-know folks are saying. Frankly, I feel it is best to play it safe — don't medicate "just in case."

### Nosema ceranae

Nosema ceranae is the other form of the disease that has only been seen in Western honey bees since 2006. It too is caused by a single-cell parasite, and it is also classified as a fungus. But here's where the similarities diverge. The symptoms of Nosema ceranae are not very clear, which makes it difficult to identify just by observation. For example, there is no feces-spotting of the hive. Only a microscopic analysis in the lab can confirm the presence of Nosema ceranae. But the devastation on the colony is significant, and some feel it may be a contributor to colony collapse disorder (see Chapter 11 for more on CCD).

Because information on Nosema ceranae is relatively new and treatment protocols are still evolving, it's a great idea to subscribe to a bee journal and keep yourself informed regarding the lastest news on Nosema.

## A handy chart

Table 12-1 gives you a quick overview of the big-six bee diseases, their causes, and their distinguishable symptoms. It contains a description of a healthy bee colony for comparison purposes.

**TABLE 12-1**     Honey-Bee Health at a Glance

| Situation/ Disease | What Causes It? | Appearance of Brood | Appearance of Brood Cappings | Appearance of Dead Larvae | Color and Consistency of Larvae | How Does It Smell? |
|---|---|---|---|---|---|---|
| Normal, healthy brood and bees | Terrific beekeeping! | Tight pattern of sealed and open brood cells | Light tan, brown color; slightly convex; no pinholes | No dead larvae | Plump, bright white, wet, pearly appearance | Fresh, sweet smell (or no smell at all) |
| American foulbrood (AFB) | A bacterium (spore-forming) | Scattered, spotty brood pattern | Sunken, perforated, discolored, greasy appearance | Flat and fluid-like on bottom of cell; tongue extended to the roof of the cell | Brown, dull, sticky, and ropy | Unpleasant, sharp, foul smell |
| Chalkbrood | A fungus | Scattered, spotty brood pattern | Sunken, perforated, discolored | Most often in sealed or perforated cells | White and moldy; later, white, gray, or black; hard and chalk-like | Normal |
| European foulbrood (early stages) | A bacterium | Scattered, spotty brood pattern | Some discolored, sunken, perforated | In unsealed cells, in twisted positions | Yellowish, tan or brown | Sour |
| European foulbrood (advanced stage) | A bacterium | Scattered, spotty brood pattern | Discolored, sunken, perforated | In unsealed and sealed cells, in twisted positions | Brown, but not ropy or sticky | Sour |
| Sacbrood | A virus | Scattered brood pattern; many unsealed cells | Often dark and sunken; many perforated | Most often with head raised | Grayish to black; skin has a watery, sack-like appearance | Sour or no smell |

*(continued)*

**TABLE 12-1** *(continued)*

| Situation/ Disease | What Causes It? | Appearance of Brood | Appearance of Brood Cappings | Appearance of Dead Larvae | Color and Consistency of Larvae | How Does It Smell? |
|---|---|---|---|---|---|---|
| Stonebrood | A fungus | Affected brood are usually white but can sometimes have a greenish, moldy appearance | Some cappings are perforated and covered with a greenish mold | In unsealed and sealed cells | Green-yellow or white; hard and shrunken | Moldy |
| Nosema apis | A fungus caused by a parasite | | Only affects adult bees | | | No smell |
| Nosema ceranae | A fungus caused by a parasite | | Only affects adult bees | | | No smell |

Chapter **13**

# Heading Off Honey-Bee Pests

E ven healthy bee colonies can run into trouble every now and then. Critters (two-, four-, and multi-legged) can create problems for your hives. Anticipating such trouble can head off disaster. And if any of these pests get the better of your colony, you'll need to know what steps to take to prevent things from getting worse.

In this chapter, I introduce you to a few of the most common pests of the honey-bee and what you must do to prevent catastrophe.

## Parasitic Problems

This list starts with two little mites that have gotten a lot of publicity in recent years about big problems they've created for honey bees: the Varroa mite and the tracheal mite. These two parasites in particular have become unwelcome facts of life for beekeepers, changing the way we care for our bees. You need to be aware of these pests and find out how to control them. Doing *nothing* to protect your bees from mites is like playing a game of Russian roulette.

Even those beekeepers practicing a *natural* approach to honey-bee management need to be attentive to the welfare of their colonies. Sometimes, the judicial use of certain additives is the only surefire way to eliminate problems that are harming the health of your bees.

Following package directions precisely is very important with any of the products listed in this chapter. The use of protective gloves is also recommended.

## Varroa mites

Somehow this little pest *(Varroa destructor)*, shown in Figure 13-1, has made its way from Asia to all parts of the world. Varroa has been a problem since the late 1980s (maybe longer) and has created quite a problem for beekeepers. Resembling a small tick, this mite is about the size of a pinhead and is visible to the naked eye. Like a tick, the adult female mite attaches herself to a bee and feeds on its blood *(hemolymph fluid)*.

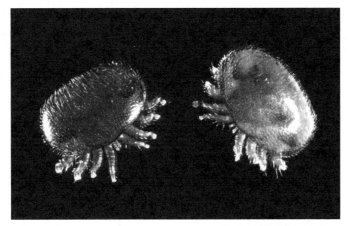

**FIGURE 13-1:**
Varroa mites can seriously weaken a hive by attaching to bees and feeding on their hemolymph (blood).

*Courtesy of USDA-ARS, Scott Bauer*

Hemolymph fluid is the "blood" of arthropods. It is the fluid that circulates in the body cavity of an insect.

Mites attached to foraging worker bees enable the infestation to spread from one hive to another. The Varroa mite is strongly attracted to the scent of drone larvae, but it also invades other brood cells just before they're capped over by the bees. Within the cells, Varroa mites feed on the developing bees and lay eggs. They reproduce at a fantastic rate and cause a great deal of stress to the colony. The health of the colony can weaken to a point that bees become highly susceptible to viruses. Within a couple of seasons, the entire colony can be wiped out.

## Recognizing Varroa mite symptoms

How do you know if your colony has a serious infestation of Varroa mites? Following is a list of some Varroa mite symptoms. If you suspect a Varroa infestation, confirm your diagnosis using one of the surefire detection techniques I describe in the next section. For starters, consider these questions:

>> **Do you see brown or reddish spots on the white larvae?** You may be seeing mites.

>> **Are any of the newly emerged bees badly deformed?** You may notice some bees with stunted abdomens and deformed wings.

>> **Do you actually *see* Varroa on adult bees?** They're usually found behind the head or nestled between the bee's abdominal segments.

**WARNING**

Finding mites on adult bees indicates a *heavy* infestation. The mites head for bee larvae first (before the larvae are capped and develop into pupae). They then feed on capped pupae. It doesn't take much to figure out that by the time the mites are prevalent on adult bees, the mite population is quite high.

>> **Did your colony suddenly die in late autumn?** Oops! You're way too late to solve the problem this year. You'll have to start fresh with a new colony next spring.

## Using two surefire detection techniques for Varroa

If you suspect a Varroa mite problem, then, by all means, confirm your diagnosis by using either the powdered-sugar-shake method or the drone-brood-inspection method. But performing one of these detection techniques before you suspect a problem is best. Varroa detection needs to be a routine part of your inspection schedule. I suggest using the powdered-sugar-shake method monthly, starting in early summer and continuing until the onset of cold weather in the autumn.

### POWDERED-SUGAR-SHAKE METHOD

**ALL NATURAL**

The powdered-sugar-shake technique for detecting Varroa is effective, natural, and nondestructive (no bees are killed in the process). You can use this process in the early spring or later fall (when no honey supers are on the hives) as well as in summer (when supers are in place). Follow these steps:

1. **Obtain a 1-pint, wide-mouthed glass jar (the kind mayonnaise comes in) and modify the lid so it has a coarse screen insert. Just cut out the center of the lid and tape or glue a wire screen over the opening (see Figure 13-2).**

   Size #8 hardware cloth (eight wires to the inch) works well. Now you have something resembling a jumbo saltshaker.

**FIGURE 13-2:**
This jar's lid has been modified for a powdered-sugar-shake mite inspection.

*Courtesy of Howland Blackiston*

2. **Scoop up about half a cup of live bees (about 200 to 300) from the brood nest and place them in the jar. Be careful that you don't scoop up the queen!**

3. **Put 3 to 4 tablespoons of powdered sugar (confectioners' sugar) into the jar and then screw on the perforated lid. (Alternatively you can use granulated sugar.)**

4. **Cover the screened lid with one hand (to keep the sugar from spilling out) and shake the jar vigorously for 30 to 60 seconds (like a bartender making a martini . . . shaken, not stirred).**

**TIP**

Shake your sugar cocktail authoritatively. Doing so dislodges any mites that are on the bees.

This action doesn't really harm the bees, but it sure wakes them up!

5. **Place the jar in a sunny spot for about a minute to allow mites to drop off the bees; then give the jar another shake like before.**

6. **Shake all the sugar through the screened top and onto a white sheet of paper.**

7. **Open the top and deposit the bees onto the entry area of the colony (they are too covered with sugar to fly home). You may want to stand to the side because they will be rather unhappy. Shake any remaining sugar onto the paper.**

8. **Count the mites, which can easily be spotted as they contrast with the white paper and white powdered sugar.**

   If you count ten or more mites, you should proceed with the recommended treatment (see the section "Knowing how to control Varroa mite problems" later in this chapter). Seeing many dozens of mites means the infestation has become significant. Take remedial action fast!

**WARNING**

Bees can be returned unharmed to the hive using this technique. Although they may be coated with sugar, their sisters nevertheless have a grand time licking them clean. Just wait 10 to 15 minutes to let them calm down before releasing them. All that jostling can make them understandably irritable — and revengeful.

### DRONE-BROOD-INSPECTION METHOD

Regrettably, the drone-brood-inspection method kills some of the drone brood. I prefer the sugar-shake method for that reason alone. If you choose this drone-brood-inspection technique, follow these steps:

1. **Find a frame with a large patch of capped *drone* brood.**

   They are the larger capped brood with distinctly dome-shaped cappings. Shake all the bees off the frame and move to an area away from the hive where you can work undisturbed.

2. **Using an uncapping fork (see Chapter 16 for an image of an uncapping fork), slide the prongs along the cappings, spearing the top third of the cappings and impaling the drone pupae as you shovel across the frame.**

3. **Pull the drone pupae straight out of their cells.**

   Any mites are clearly visible against the white pupae. Repeat the process to take a larger sampling. See Figure 13-3.

Two or more mites on a single pupa indicate a serious, heavy infestation. Two or three mites per 50 pupae indicate a low to moderate infestation. But remember, whenever you see *any* mites at all, it's time to take action! (See "Knowing how to control Varroa mite problems" coming up.)

### SCREENED-BOTTOM-BOARD METHOD

The screened-bottom-board method is not as reliable an evaluation method as the previous two, but it's worth a mention. About 10 to 15 percent of Varroa mites routinely fall off the bees and drop to the bottom board. But if you use a screened bottom board (sometimes called a "sticky" board), the mites fall through the screen and onto a removable tacky white board.

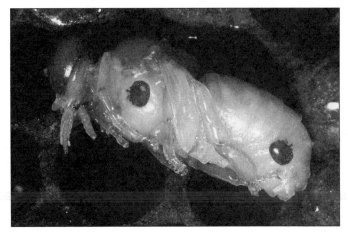

**FIGURE 13-3:**
Varroa mites first attach themselves to drone pupae, so that's a good place to look for evidence of an infestation. Can you see the mites on these pupae?

*Courtesy of USDA-ARS, Scott Bauer*

When this white board is in place, mites fall through the screen and become stuck to the sheet (you apply a thin film of petroleum jelly or cooking spray to the sheet to help the mites stick).

Just insert the sheet for 24 hours and then remove it to count the mites. If the number of mites is more than 24 in a 24-hour period, then appropriate control measures should be taken. The board can be wiped clean and readied for its next use.

A screened bottom board is also an excellent way to improve ventilation in the hive.

**TIP**

## Knowing how to control Varroa mite problems

Several techniques and products are available that help reduce or even eliminate Varroa mite populations. Some involve chemicals, and some favor an all-natural approach. Here are the ones that I suggest you consider and a few that I think you should avoid as a new beekeeper.

### GO AU NATUREL

You don't always have to use chemicals to deal with Varroa mites. Integrated pest management (IPM) is the practice of controlling honey-bee pests with the minimal use of chemicals. (See Chapter 11 for more on IPM.) As it relates to controlling Varroa mites, here are a couple of nonchemical techniques to consider. Both are fairly labor intensive — but hey, no harsh chemicals are used!

**ALL NATURAL**

>> **Use drone comb to capture Varroa mites.** Bee suppliers sell a special drone foundation that has larger hexagons imprinted in the sheet. The bees will

build drone comb only on these sheets. That's useful because Varroa mites prefer drone brood over worker brood. By placing a frame of drone comb in each of your hives, you can capture and remove many mites. After the drone cells are capped, remove the frame and place it overnight in your freezer. This kills the drone brood and the mites that have invaded the cells. Then uncap the cells and place the frame (with the dead drone brood and dead mites) back in the hive. The bees will clean it out, removing the dead drone brood and mites. The cells will get filled again, and you repeat the process.

» **Powdered-sugar dusting to control Varroa mites.** This involves dusting the bees with powdered sugar. (Note it's best to find a powdered sugar without added cornstarch, although some claim this is not so critical. Play it safe and ask your bee suppliers for a "pure" powdered sugar.) The idea behind this technique is that the powdered sugar knocks many of the mites off the bees, and the mites fall down through the screened bottom board and perish in the grass below the hive (this assumes you are using an elevated hive stand and a screened bottom board with the insert removed). Use this method continually during the season to help keep the mite populations below harmful levels.

Here's the process:

1. **Sift a pound of powdered sugar using a baking flour sifter.**

   Do this twice to ensure no lumps. This should be done on a day with low humidity.

2. **Put the sifted sugar into an empty (and cleaned) baby-powder container. (Alternatively, you can improvise your own container.)**

3. **Smoke and open the hive. Then place a frame on the rest and do your dusting thing.**

   Remove frames one by one, and dust the bees with the sugar. A frame rest comes in handy here (see Chapter 5 for an image of a frame rest).

   The key operating word is *dusting* the bees . . . not coating them with loads of powdered sugar. You want to master a technique that makes light clouds of sugar dust — don't shake the sugar directly on the bees.

4. **Avoid dusting any open cells.**

   You just want to dust the backs of the bees.

5. **Put the dusted frame back into the hive and repeat this process with each frame.**

6. **When done, put a little extra dusting along all the top bars.**

This should be repeated once a week for two to three weeks.

## SYNTHETIC CHEMICAL OPTIONS

There are a few effective and approved *synthetic miticide chemicals* (chemicals that kill mites). One is fluvalinate, which is sold under the brand name *Apistan* and is available from your beekeeping supplier. Another is amitraz, sold under the brand name *Apivar,* and yet another is coumaphos (marketed as *CheckMite+*). Formic acid is also used as a treatment for tracheal mites (sold in plastic strips under the brand name *Mite Away Quick Strips*). In addition there are "soft" (safer) chemicals such as thymol (marketed as *Apiguard*). Also consider oxalic acid (which sounds scary but it's actually derived from plants). It's a bit tricky to use and should only be used when there is little or no brood in the hive. When any of the detection techniques mentioned earlier in this chapter indicate Varroa mites, consider using one of these treatments by carefully following the directions on the package.

**WARNING**

>> **Apistan (fluvalinate):** Apistan is packaged as chemical-impregnated strips that look kind of like bookmarks. Hang two of the plastic strips in the brood chamber between the second and third frames and the seventh and eighth frames. You're positioning the strips close to the brood so the bees naturally come into contact with the pesticide they contain. The bees will brush up against each other and transfer the fluvalinate throughout the hive.

> The U.S. mite population has developed resistance to this product, and treatment is not considered reliable.

>> **Apivar (amitraz):** Apivar is packaged as chemical-impregnated strips that are hung in the hive. The principle behind its use is similar to that of Apistan. The chemical (pesticide) used is the main difference between the two products. The mites have not yet developed resistance to this chemical treatment, but you should keep abreast of developments as resistance may occur in coming years.

>> **CheckMite+:** CheckMite+ is a product manufactured by the Bayer Corporation (of aspirin fame). Like Apistan, it also consists of strips impregnated with a chemical pesticide. But in the case of CheckMite+, the chemical is *coumaphos* — an ingredient used in deadly nerve gas. It's tricky to use. And mite resistance to this chemical is widespread. However, this is the best chemical currently on the market for small hive beetle control.

>> **Mite Away Quick Strips (formic acid):** Formic acid is considered a *natural-source* chemical. The Mite Away Quick Strips formulation consists of strips that are laid across the top bars of the brood area. The formic acid is highly effective in penetrating the brood cappings to kill mites within cells. It's a seven-day treatment (other products require much longer treatment periods). Note that the manufacturer states that this product can be used when there is honey on the hive for human consumption.

» **Apiguard (thymol):** Apiguard is a natural product specifically designed for use in beehives. It is a slow-release gel matrix, ensuring correct dosage of the active ingredient thymol. Thymol is a naturally occurring substance derived from the plant thyme, and for this reason it can be considered a more natural product for use with your colonies. It is effective against the Varroa mite and is also active against both tracheal mite and chalkbrood. It is easy to use and much safer than formic acid or coumaphos.

TIP

Because Varroa mites can develop a resistance to these medications, it is prudent to alternate between two or more of them from one season to the next.

TIP

Check the product label to see if the treatment you are using is safe to use when there is honey on the hive for human consumption (some are okay to use, some are not). To play it safe, don't treat your bees with medication when you have honey supers on the hive. If you do, your honey might become contaminated and cannot be used for human consumption. *Note:* Honey exposed to any of these medications can be safely fed to your bees, however.

## Tracheal mites

Another mite that can create serious trouble for your bees is the tracheal mite (*Acarapis woodi*), shown in Figure 13-4. These little pests are much smaller than the period at the end of this sentence and can't be seen with the naked eye. Dissecting an adult bee and examining its trachea under magnification is the only way to identify a tracheal mite infestation.

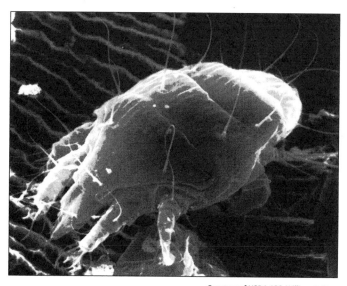

**FIGURE 13-4:**
An adult
tracheal mite
(*Acarapis woodi*).

Courtesy of USDA-ARS, William E. Styer

As its name implies, this mite lives most of its life within the bee's trachea (breathing tubes), as shown in Figure 13-5. Mated female mites pass from one bee to another when the bees come in close contact with each other. Once the mite finds a newly emerged bee, she attaches to the young host and enters its tracheal tubes through one of the bee's spiracles — external opening holes that are part of the respiratory system. Within the trachea the mite lays eggs and raises a new generation. The tracheal mite causes what once was referred to as *acarine disease* of the honey-bee (a rather old-fashioned term not used much these days).

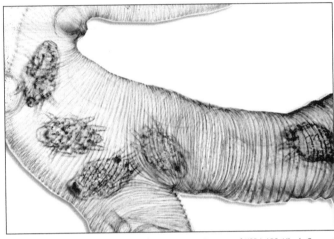

**FIGURE 13-5:** Tracheal mites (seen in this magnified photo of an infected bee's tracheal tubes) pose a serious threat to the health of bee colonies.

*Courtesy of USDA-ARS, Lila de Guzman*

In my opinion, this mite causes more trouble for hobbyist beekeepers than Varroa. Early detection of bad infestations is difficult. As a result, tracheal mites can lead to the total loss of a colony before you're even aware that your bees are infested. Infestations are at their worst during winter months when bees are less active. Her majesty isn't laying eggs, so no new bees are emerging to make up for attrition. Winter also is when beekeepers don't routinely inspect the colony. Thus, seemingly healthy colonies with plenty of food sometimes suddenly die during late winter or early spring.

## Symptoms that may indicate tracheal mites

The only surefire way to detect tracheal mites involves dissecting a bee under a microscope — a little tricky for the novice, and not everyone has a dissection microscope in the hall closet. Whenever you suspect tracheal mites, call your state apiary inspector (if one is available) or check with local bee clubs about labs that are doing this diagnosis.

A few clues may indicate the presence of tracheal mites. But the symptoms, listed as follows, are unreliable because they also may indicate other problems:

>> **You see many weak bees stumbling around on the ground in front of the hive.** (This condition could also be an indication of Nosema disease; see Chapter 12.)

>> **You spot some bees climbing up a stalk of grass to fly, but instead they just fall to the ground.** This happens because mites clog the trachea and deprive the bee of oxygen to its wing muscles.

>> **You notice bees with *K-wings* (wings extended at odd K-shaped angles and partly opened — not folded in the normal closed position).** This also can be an indication of Nosema disease.

>> **Bees abandon the hive (abscond) in early spring despite ample honey supplies.** This can happen even late in the fall when it's too late to remedy the situation.

## How to control tracheal mite problems

Tracheal mite infestations are a problem, not a hopeless fate. You can take steps by using several techniques that I list in the following sections to prevent things from getting out of control. It isn't a case of just one technique working well. Play it safe by using a combination of some or all of these methods.

## Natural source options

Several remedies for tracheal mites involve a more natural approach than using synthetic chemicals.

### SUGAR-AND-GREASE PATTIES

Less popular in recent years than it was a decade ago, placing patties of sugar and grease in the hive is a natural treatment for tracheal mites that some beekeepers use year-round (even during the honey harvest season — unless you are adding the wintergreen oil option). As the bees feed on the sugar, they become coated with grease. The grease impairs the mite's ability to reproduce or latch onto the bees' hairs. Whatever the scientific reason, the treatment seems to work pretty well and is a natural defense against tracheal mites.

Place one patty on the top bars of the brood chamber, flattening out the patty as needed to provide clearance for the inner cover and replacing it as the bees consume it. One patty should last a month or more.

Here's a recipe I've used for sugar-and-grease patties:

1½ pounds of solid vegetable shortening (such as Crisco)

4 pounds of granulated sugar

½ pound honey

**Optional:** Add a cup of mineral salt (the orange/brown salt available at farm supply stores — it's used to feed livestock). Pulverize the salt in a blender, breaking it into a fine consistency. The bees seem to like it, and this in turn encourages better interaction with the sugar and grease patties.

Mix all these ingredients together until smooth. Form into about a dozen hamburger-size patties. Unused patties can be stored in a resealable plastic food bag and kept frozen until ready to use.

*Note:* As an option, you may add 45 milliliters (1.5 ounces) of natural food-grade wintergreen oil to the mixture, provided that you're not using this treatment while honey for human consumption is on the hive. Some beekeepers have indicated enhanced results when adding essential oils.

WARNING

Treating with sugar-and-grease patties can sometimes be an attractive lure for the small hive beetle, so be on the lookout for these unwelcome critters. There is more about the hive beetle later in this chapter.

## MENTHOL CRYSTALS

Menthol crystals are the same ingredient found in candies and cough drops. Menthol is derived from a plant, making it a natural alternative to chemical pesticides. Prepackaged bags containing 1.8 ounces of menthol crystals are available from your beekeeping supplier.

WARNING

Not all menthol products are approved for use with honey bees. Be sure to purchase menthol packets (such a Mite-a-Thol) from reputable beekeeping supply vendors.

Place a single packet on the top bars of the brood chamber toward the rear of the hive (see Figure 13-6). Setting the packet on a small piece of aluminum foil prevents the bees from chewing holes in the bag and carrying away the menthol. Bees are tidy and try their best to remove anything they don't think belongs in the hive. Leave the menthol in the hive for 28 *consecutive* days when the outdoor temperature is between 60 and 80 degrees Fahrenheit (16 and 27 degrees Celsius). The menthol vapors are effective only at these temperatures. That means the product is *temperature dependent* — you can only use it when the weather is warm. Treat for tracheal mites in both the late spring and early autumn.

Courtesy of Mann Lake Ltd.

FIGURE 13-6:
In warm weather a packet of menthol, such as Mite-a-Thol, placed on the top bars of the brood chamber helps control tracheal mites.

**REMEMBER**

Honey for human consumption must be taken off the hive whenever *any* medications are used (natural or otherwise). You can safely add honey supers three to four weeks after menthol is removed from the hive.

### ESSENTIAL OILS

Several studies have tested the effectiveness of using essential oils as a means of controlling mite populations. Essential oils are those natural extracts derived from aromatic plants such as wintergreen, spearmint, lemon grass, and so on. These oils are available from health-food stores and companies that sell products for making soap.

Pioneering work on the use of essential oils in honey-bee hives has been conducted by Bob Noel and Dr. Jim Amrine (West Virginia University).

For information and links to sites devoted to Varroa and tracheal mite studies, visit Dr. Amrine's website at www.wvu.edu/~agexten/Varroa.

## Synthetic chemical options

When using any of these products, be certain to follow the instructions very carefully.

>> **Apiguard (Thymol):** Apiguard is mentioned earlier in this chapter. In addition to being effective against the Varroa mite, it is also active against both tracheal mite and chalkbrood. It is easy to use, but as always, follow directions precisely.

# ALL-NATURAL FOOD SUPPLEMENTS

One of the original all-natural products on the market is Honey B Healthy. It contains pure essential oils (spearmint and lemon grass oils) and is sold as a concentrated food supplement (see the figure that follows) that's added to the sugar syrup you feed your bees in the spring and autumn. It was developed by Bob Noel and Jim Amrine, pioneers in the use of essential oils with honey bees. The manufacturers of this and similar products make no claims about their ability to kill mites, but some beekeepers feel this product keeps bees healthy and strong even in the presence of Varroa and tracheal mites. Since Honey B Healthy was first introduced, a whole bunch of similar products have been entering the market. Check with your favorite beekeeping supplier for a complete list of these food supplements.

*Courtesy of Howland Blackiston*

>> **Api-Life VAR (Thymol):** Api-Life VAR is 95 percent effective while leaving the least residue of any product available to beekeepers. It contains thymol plus three essential oils. Each colony is treated by breaking a wafer into four pieces and placing the pieces around the brood nest. This product is not approved for use in all areas, so be sure to check its availability with your beekeeping supplier.

>> **Apivar:** Apivar, mentioned earlier for Varroa mite control, is also effective in controlling tracheal mites.

>> **Mite Away Quick Strips (formic acid):** Mentioned earlier in this chapter as a treatment option for Varroa mites, these products are also effective for controlling tracheal mites.

## Zombie (Phonid) flies

In 2012 this critter became a problem for honey bees in some parts of the United States (as if there were not enough problems already). And it became a bonanza for morbid journalists seeking sensational headlines. Here's why.

The fly (*Apocephalus borealis*, to be precise) deposits eggs into the body of an adult bee, where the larva feeds and develops, eventually emerging from the bee to pupate, and sometimes decapitating the bee in the process. Horrors. While infected, the bees begin to act more and more disoriented, flying aimlessly from the hive at night (bees never fly at night), stumbling and crawling about the ground, and eventually dying. The term *zombees* was the resulting headline. Given the public's craze for all things zombie related, it's no wonder the media has had a field day with this story.

So while it's not yet as much of a threat as the other parasitic problems listed in this chapter, it's worth a mention. The best defense is maintaining a strong and robust colony.

# Other Unwelcome Pests

As if the mites and flies already mentioned weren't enough, there are other types of pests that can make life miserable for you and your bees.

## Wax moths

Wax moths can do large-scale damage in a weak hive (see the color-photo section of this book). They destroy the wax comb, which ruins the colony's ability to raise brood and store food. But they don't usually become a problem in a strong and healthy hive because bees continually patrol the hive and remove any wax moth larvae they find. Therefore, if you see wax moths, you probably have a weak colony. Keeping your bees healthy and reducing hive boxes and frames to correspond to colony size are the best defenses against wax moths. Once an infestation overtakes a weak colony of bees, there is nothing you can do to get rid of it. It's too late.

The story is different when comb is stored for the winter. With no bees to protect these combs, the wax is highly susceptible to invasion by wax moths. But in this case, steps can be taken to keep the moths from destroying the combs over the winter. The use of PDB crystals (para dichlorobenzene, or Para Moth, as shown in Figure 13-7) on stored supers and hive bodies can kill the moths and larvae that would otherwise destroy the wax.

**WARNING**

This product is not for use on supers that have contained or will contain honey for human consumption. Follow product instructions carefully.

**ALL NATURAL**

Here's a chemical-free approach that I urge you to use. You can destroy wax moth larvae by placing the frames in the deep-freezer for 24 hours. If you don't have a big freezer, you can freeze two or three frames at a time. After they've had the freeze treatment, put the frames back in the supers and store them in tightly sealed plastic garbage bags. The colder the storage area, the better.

## Small hive beetle

More bad news for bees came in 1998 when the small hive beetle was discovered in Florida. Most common beetles that wander in and out of a hive are not a problem, so don't panic if you see some bugs. But the small hive beetle, originally from Africa, is an exception. The larvae of this beetle eat wax, pollen, honey, bee brood, and eggs. In other words, they gobble up nearly everything in sight. The beetles also — yuck! — defecate in the bees' honey, causing it to ferment and ooze out of the comb (see the image in the color-photo section). Things can get so slimy and nasty that the entire colony may pack up and leave. Who can blame them?

### Determining whether you have a small hive beetle problem

Be on the lookout for little black or dark brown beetles (see Figure 13-8) scurrying across combs or along the inner cover and large numbers of adult beetles on the bottom board. You may even notice their creamy larvae on the combs and bottom board. If you see some suspects, don't panic; get them identified. There are many lookalikes.

**FIGURE 13-8:**
The small hive beetle has become a significant problem for beekeepers in some states (mostly in the Southeast).

## How to control the small hive beetle

First of all, keeping your colonies strong and healthy is your best natural defense. In addition, you need to destroy any small beetles you see during routine inspections. If infestation levels appear heavy, medicating your hive may be necessary. Presently, two approved treatments for the small hive beetle are sold under the brand names of CheckMite+ to kill the adult beetles inside the hive and GardStar for soil application in the apiary to halt their reproduction.

TIP

If you suspect that you have the small hive beetle, contact your state apiary inspector (if you have one) or consult with members of your regional bee club. It's important that you do your part to keep this new pest from spreading across the country. The inspector or fellow club members will let you know what kinds of treatments are effective in your area. They might even help you with the treatment!

## SMALL HIVE BEETLE TRAP

**ALL NATURAL**

In the spirit of natural hive management, here's a nonchemical means for controlling the population of small hive beetles. The small hive beetle trap is a two-piece plastic trap that sits on the existing bottom board. There are perforations in the top piece, and the bottom tray is filled with vegetable oil. A wooden shim provides the proper spacing at the entrance. The bees will chase the beetles as they enter the hive, causing the beetles to seek a hiding spot. As they retreat through the holes of the trap, they fall into the oil and drown.

# Ants, ants, and more ants

Ants can be a nuisance to bees. A few ants here and there are normal. A healthy colony keeps the ant population under control. But every now and then things can get out of hand, particularly when the hive is too young or too weak to control the ant population. Sometimes simply more ants are around than the colony can handle. When ants overrun a colony, the bees may *abscond* (leave the hive). But you can take steps to control the ant population *before* it becomes a crisis. Two natural approaches you can use if you notice more than a few dozen ants in the hive are the following:

>> **Send cinnamon to the rescue:** Purchase a large container of ground cinnamon from a restaurant supply company. Sprinkle the cinnamon liberally on the ground around the hive. Sprinkle some on the inner cover. Your hive will smell like a giant breakfast doughnut. Yummy! The bees don't mind, but the ants don't like it and stay away. Remember to reapply the spice after the rain washes it away.

>> **Create a moat of oil:** This technique is a useful defense against ants. You'll need a hive stand with legs. (This is a good idea even if you don't have an ant problem, because raising the hive off the ground is a back-saver for you!) Place each of the stand's four legs in a tin can — old tuna cans or coffee cans are fine. Fill the cans with oil (such as canola, vegetable, mineral, and so on). You can use water, but it will evaporate quickly whereas the oil will not. The ants won't be able to cross the "moat" and thus are unable to crawl up into the hive (see Figure 13-9).

# Bear alert!

Do bears like honey? Indeed they do! And they simply crave the sweet honey-bee brood. (I've never tried it myself, but I suspect it's sweet.) If bears are active in your area (they're in many states within the continental United States), taking steps to protect your hive from these lumbering marauders is a necessity. If they catch a whiff of your hive, they can do spectacular and heartbreaking damage, smashing apart the hive and scattering frames and supers far and wide (see Figure 13-10). What a tragedy to lose your bees in such a violent way. Worse yet, you can be certain that once they've discovered your bees, they'll be back, hoping for a second helping.

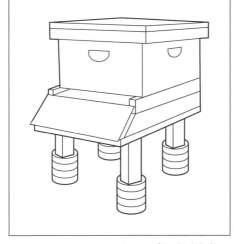

**FIGURE 13-9:** Placing the legs of your elevated hive stand in cans of oil prevents ants from marching into your hive.

*Courtesy of Howland Blackiston*

**FIGURE 13-10:** These beehives were shattered to smithereens by a hungry bear.

*Courtesy of* Bee Culture Magazine

TIP

The only effective defense against these huge beasts is installing an electric fence around your apiary. Anything short of this just won't do the trick.

TIP

If you're ever unlucky enough to lose your bees to bears, be sure to contact your state or local conservation department. You may qualify for remuneration for the loss of your bees. And the department may provide financial assistance for the installation of an electric fence. They may even offer a service for trapping and relocating the offender. However, with decreasing state budgets, these perks may be increasingly difficult to come by.

## Raccoons and skunks

Raccoons are clever animals. They easily figure out how to remove the hive's top to get at the tasty treats inside. Placing a heavy rock on the hive's outer cover is a simple solution to a pesky raccoon problem.

Skunks are insect eaters by nature. When they find insects that have a sweet drop of honey in the center . . . bonanza! Skunks and their families visit the hive at night and scratch at the entrance until bees come out to investigate. When they do, they're snatched up by the skunk and . . . gulp! Skunks can put away quite a few bees during an evening's banquet. In time, they can decimate your colony. These raids also make your bees decidedly more irritable and difficult to work with. You need to put a quick end to skunk invasions.

Putting your hive on an elevated stand is an effective solution for skunk invasions. The skunk then must stand on his hind legs to reach the hive's entrance. That exposes his tender underbelly to the bees — and have no doubts, the bees know what to do next!

You may wonder how the bees feel about skunk scent and whether it bothers them. I can't really give you a good answer to this: My bees have never told me how they feel about the smell, nor do I think I've ever known a skunk to spray a hive. But my dog has some stories to tell!

Another solution is hammering a bunch of nails through a plank of plywood (about 2-feet square) and placing it on the ground in front of the hive with the nail points sticking up like a bed of nails. Or you can use carpet tack strips available at home-improvement stores. No more skunks. Just be sure *you* remember the plank's there when you go stomping around the hive!

## Keeping out Mrs. Mouse

When the nighttime weather starts turning colder in early autumn, mice start looking for appropriate winter nesting sites. A toasty warm hive is a desirable option. The mouse may briefly visit the hive on a cool night when bees are in a loose cluster. During these exploratory visits, the mouse marks the hive with urine so she can find it later. When winter draws nearer, the mouse returns to the marked hive and builds her nest for the winter.

I can assure you that you don't want this to happen. Mice do extensive damage in a hive during the winter. They don't directly harm the bees, but they destroy comb and foundation and generally make a big mess. They usually leave the hive in early spring, long before the bees break winter cluster and chase them out or sting them to death. Nesting mice isn't the surprise you want to discover during your early spring inspection. Anticipate mouse problems and take these simple steps to prevent mice from taking up winter residence in your hive:

1. **As part of winterizing your hive, use a long stick or a wire coat hanger to "sweep" the floor of the bottom board, making sure no mouse already has taken up residence.**

   Shoo them out if they have.

2. **When you're sure your furry friends are *not* at home, secure a metal *mouse guard* along the entrance of the hive (see Figure 13-11).**

   This metal device enables bees to come and go and provides ample winter ventilation, but the mouse guard's openings are too small for Mrs. Mouse to slip through.

**FIGURE 13-11:** Installing a metal mouse guard prevents mice from nesting in your hive during winter.

*Courtesy of Howland Blackiston*

**TIP**

The early use of mouse guards pays dividends. Mama mouse makes early visits to the hive (when the weather is still mild), looking for a suitable home for the winter. She will "scent" the inside of the hive at night when the bees are inactive and then will return in the cooler weather to take up winter residence. So don't delay getting those guards on *before* she makes her rounds.

Using a wooden entrance reducer as a mouse guard doesn't always work. The mouse nibbles away at the wood and makes the opening just big enough to slip through.

## Some birds have a taste for bees

If you think you notice birds swooping at your bees and eating them, you may be right. Some birds have a taste for bees and gobble them up as bees fly in and out of the hive. But don't be alarmed. The number of bees that you'll lose to birds probably is modest compared to the hive's total population. No action need be taken. You're just witnessing nature's balancing act.

# Pest Control at a Glance

Table 13-1 provides treatment options for the various pests mentioned in this chapter.

**TABLE 13-1**    ## Solutions to Common Pest Problems

| Pest Problem | Solution(s) | Comments |
|---|---|---|
| Varroa mites | Use drone comb | The drone-comb and powdered-sugar-dusting methods use a natural solution |
|  | Powdered-sugar dusting |  |
|  | Apistan (fluvalinate) | Alternate your medication treatments to prevent the mites from building a resistance to the pesticide |
|  | Apivar |  |
|  | CheckMite+ | Apiguard uses natural (nonharsh) ingredients |
|  | Mite Away Strips (formic acid) |  |
|  | Apiguard (thymol) |  |
| Tracheal mites | Menthol crystals | All of these options use natural (nonharsh) ingredients |
|  | Sugar-and-grease patties |  |
|  | Apiguard (thymol) |  |
|  | Mite Away Strips (formic acid) |  |
|  | Api-Life VAR |  |
|  | Apivar |  |

| Pest Problem | Solution(s) | Comments |
|---|---|---|
| Zombie flies | Maintain strong, healthy hives | Also called zombees |
| Wax moths | PDB crystals (para dichlorobenzene)<br><br>Freezing frames | Applied only to empty supers and hive bodies where there is comb that was used to raise brood<br><br>Freezing is a natural, nonchemical solution |
| Small hive beetles | Check-Mite+ (coumaphos)<br><br>GardStar<br><br>Hive trap | Harsh chemical<br><br>GardStar is applied to the ground surrounding the hive<br><br>The hive trap is a natural, nonchemical solution |
| Ants | Cinnamon<br><br>Oil or water moat | Both the cinnamon and oil-moat methods are natural, nonchemical solutions |
| Bears | Electric fence around your hives | Some states may offer financial compensation for an electric bear-fence installation |
| Raccoons and skunks | Elevated hive stand<br><br>Bed of nails | Elevating a hive is also easy on your back!<br><br>Watch where you step! |
| Mice | Metal mouse guard placed at hive entrance | Install guards well before the cold weather approaches |

# Chapter **14**

# Raising Your Own Queens

S pecial thanks to my friend and EAS Master Beekeeper Leslie Huston for her help preparing this chapter.

In this chapter, I introduce you to different methods for raising your own queens. The process can be fairly involved, but after you're familiar with some basic concepts, you'll be on your way to success.

Several books have been written on this topic alone, and I recommend you do some further reading if you want to pursue this fascinating component of beekeeping. I cover some of the fundamentals and a few easy methods to get your feet wet.

## Why Raising Queens Is the Bee's Knees

Some colonies are particularly delightful to work with. Nice temperament, healthy bees, great honey producers, and resilient enough to survive winter. This is the kind of colony that's a pleasure to be around. It's the queen who possesses the genetics that provide the colony with these desirable traits.

When you raise your own queens, you can avoid "imported" problems. Every time you bring bees into your apiary from other sources, you run the risk of bringing unwanted "hitchhikers" along — like the small hive beetle, American foulbrood, and Africanized honey bees, just to name a few. When established, these troubles can be difficult to get rid of.

Raising healthy, well-developed queens with genetics that are tailored to your regional climate will produce healthy colonies and can help you avoid the multitude of worries and problems currently facing honey bees (including colony collapse disorder, or CCD — see Chapter 11). Robust colonies are resistant to pests, chemicals, and diseases and can better handle the weather in your area.

Raising superior queens is possible for the small-scale beekeeper. With a little attention and by providing the bees with the things they need, you can raise strong, healthy queens — right from the bees you already have.

And here's the best part. Raising queens is fun! You'll experience a whole new dimension of these fascinating insects. And you'll feel proud when you hold your homegrown queen in your hand.

Sure, you will encounter some challenges, but they will help you grow as a beekeeper. Queen rearing will put you in better touch with the overall health and well-being of all your bees. Even if you don't make queen rearing a permanent part of your beekeeping repertoire, by trying it you'll begin to appreciate what makes a good queen and what to look for when shopping for one.

**TIP**

You can raise as few or as many queens as you like. I suggest that as a first step, you raise just enough queens for your own needs, rather than enough queens to sell to the public. Then, if you want, kick things up a notch and raise queens that you can sell to other beekeepers. Your carefully raised, local stock will likely be a big hit with other beekeepers in your neighborhood.

There is a subtle but important difference in definition of terms to be appreciated:

>> Queen *rearing* is the process of raising queen bees. There are several methods for queen rearing. I cover the general principles here and describe in some detail a couple of the most popular methods.

>> Queen *breeding* is the act of identifying and selecting queens with superior characteristics to use as the parents of subsequent generations. This selection process results in a greater tendency for descendent generations to exhibit

the traits for which you select. Using gentleness as an example: Selecting queens from gentle colonies to raise new queens results in new colonies that are gentler than if you had not done that careful selection.

In this way, queen rearing and queen breeding are separate yet related processes.

# Understanding Genetics

The study of genetics and inheritance is a big topic. Huge, really. What follows are some basic tenets relevant to queen rearing and queen breeding.

Just like people, chickens, and peas, genes determine the particulars of appearance and capabilities. In bees, there are genes that control body color, disease resistance, and temperament, and these traits are passed from one generation to the next. Which body color, what type and level of disease resistance, gentle or cranky . . . the possible outcomes for the offspring are limited to the genes carried by the parents. Honey bees are responsive to selection, and in just a few seasons you can influence the overall health of your colonies.

All queens and workers develop from fertilized eggs. They possess a full set of chromosomes and have a complete genetic makeup (a mother and father). Drones develop from unfertilized eggs and have half the full set of chromosomes (a mother, but no father). If an individual winds up with two different genes for color, the dominant color will be expressed.

## Dominant and recessive genes

In honey bees, the females are *diploid,* meaning they possess two sets of genes — one from their father and one from their mother. If they receive a different gene from each parent, the dominant gene will be expressed, and the recessive one won't be evident (see Figure 14-1).

When the queen's ovaries produce eggs, the genes are divided in half. If a queen's body-color gene pair are the same, she will exhibit that color. If they are different, she will exhibit the dominant color. Half of her eggs receive the dominant color gene, whereas the other half of her eggs carry the gene for the recessive color. If that egg is fertilized — if it goes on to become a female, not a male — then the ensuing larva will have two genes for color. If the two genes match, then that is the color the bee will be. If they are different, then again, she will show the dominant color and carry the gene for the recessive color; it's in her genes, but not evident.

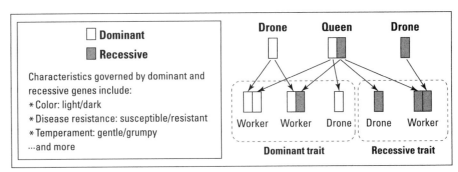

FIGURE 14-1:
This illustrates
the dominant
versus recessive
gene expression.

So as you get into queen rearing, you use the very best queen or queens in your operation as queen mothers and encourage other colonies with desirable traits to produce drones.

As you raise successive generations of queens, you can begin to select queens to continue the improvement process. In just a few seasons, it is possible to have a substantial influence on the nature of your bees, whether it is color, temperament, or disease resistance.

If you wind up with more queens than you can use, get them to good homes. You can select the very best to keep for your own colonies, and from those you can select your next queen and drone mothers. But the other queens are fine queens, too, so no harm in selling (or giving) them to fellow beekeepers. Be sure to ask for feedback from those you sell or gift to — get their impression of the health and character of the queens you've raised, and make notes so you can continually improve the genetics.

## Inbreeding versus outcrossing

Imagine if you had a truly isolated mating yard, and you selected certain traits over many generations. The gene pool of your breeding stock would become more and more focused, with less variation, and would yield more consistent bees with most or all having the traits for which you've selected. But beyond a certain point, that genetic focus can become a liability, and inbreeding can occur. Inbreeding in honey bees is evidenced by a scattered brood pattern. Rather than the nice, continuous expanse of capped worker brood, there are many *skipped cells* caused by the queen mating with too many drones that are too closely related to her. (**Note:** There may be other causes of spotty brood. Chapter 12 goes into detail about other diseases that can cause this situation.) The workers can tell this inbreeding, and they remove the inbred larva, leaving skips in the brood pattern. A skip here and there is normal, but in inbred bees, the skips become much more prevalent.

But for the small-scale queen rearer and bee breeder, this is not likely to happen. Truly isolated mating areas are rare, and it's very likely that even though you stack the deck, there will be incursions from other bee stock. Bees from neighboring beekeepers or from feral colonies will mix with your stock enough to avoid inbreeding.

If you believe you have a fairly isolated mating area, you may want to deliberately introduce some unrelated bees to your operation from time to time. Bees from different sources, different races, and other regions bring some diversity to your stock. This counteracts any possible inbreeding with a dose of outcrossing and keeps the genetic diversity from becoming too limited.

## Accentuate the positive

With bees, just like all plants and animals, traits — both good and bad — are passed from one generation to the next. With respect to livestock (and your bees can be considered livestock), it is common to select from your "herd" to retain the best traits and minimize or eliminate the worst. This is how the many different dog breeds came into existence, and it has also shaped the characteristics of our chickens, cows, corn, and so on.

To breed better bees, you need one or more mothers and a whole lot of fathers. So there are two types of colonies you will select to be your breeding stock:

>> **Queen mother colony:** Your very best colony and the queen that rules the colony is the one to use to raise more queens. This queen is called the *queen mother*.

>> **Drone mother colony:** Wherever you plan to allow your queens to do their mating, you want to have the most desirable *drone* (male bee) stock available. You don't want drones from subpar colonies contributing their poor genetics to your fine queens. Unless you live in an area where you can guarantee isolation, you will be unable to control the drone gene pool completely. But you can stack the deck in your favor by doing your best to saturate the area with lots of healthy drones from colonies with desirable traits.

Whether you are selecting a queen mother or a drone mother, you'll want to consider certain traits. Some traits are more hereditary than others. In the case of honey bees, here are some of the most desirable traits to look for when selecting colonies for your breeding project:

>> **Gentleness:** Gentleness is an important (and hereditary) trait for bees to have — no beekeeper wants to be stung. You can test a colony for gentleness by vigorously waving a wand with a black leather patch at the end over an

open hive. This will alarm the bees, and they may rush up to sting the leather patch. After a minute or so, count the stingers on the patch. The colonies with the fewest stingers in the patch are the most gentle.

>> **Resistance to disease and pests:** Bee breeders and commercial queen producers are making progress with breeding bees that are resistant or more tolerant to disease (for more on bee diseases, see Chapter 12). You can do this as well by identifying those colonies that are the most robust and require the least treatment. Raise queens using stock from these superstar colonies.

>> **Hardiness:** Winter hardiness is especially important to beekeepers in climates that have long, cold winters. If you live in a climate where winter's deep freeze lasts 10 to 12 weeks or more, then you may regard winter as the enemy, at least as far as your bees' survival goes. Winter can also be viewed as an objective selector. Colonies that survive a long cold spell must be healthy and strong. They must produce and store enough honey to fuel their winter hunker-down. And they should slow down their brood-rearing in the fall efficiently, and start up in the spring in time to have their numbers grow to take advantage of the spring nectar flow. It takes a healthy, productive, well-rounded colony to survive a northern winter. These are desirable traits for your colonies.

>> **Productivity:** Some say you can't really select for honey production when breeding queens because there are too many other factors, such as the weather (if it doesn't rain, then there are fewer flowers, and honey production will be low). So whereas honey production may be more a function of environment than genetics, if one of your colonies consistently produces more honey than another, consider that colony's queen for breeding.

TIP

No matter how you go about raising queens, provide your queen-rearing operation with every advantage. Go easy on, or better yet, avoid completely, any chemical treatments. And make sure your bees have plenty of honey and pollen.

## What Makes a Queen a Queen

All bees are essentially created equal to start. The queen develops from a fertilized egg, just the same as the workers do. After two to three days, the fertilized egg of the bee destined to be a queen hatches into a young larva, just as with worker bees. A day or so after the egg hatches into a young larva, it's decision time. A larva that goes down the road to queendom continues to receive a plentiful supply of rich royal jelly, and only royal jelly. But larvae that will become workers are switched to a nourishing but coarser diet of honey and pollen. See Chapter 2 for more about the life cycle of bees.

A queen takes just 16 days to develop. (A worker takes 21, and a drone takes 24.)

Here's how the queen's development proceeds and a few notes for your queen-rearing efforts:

>> **Days 1–2:** The egg stands on its end on the bottom of the cell.

>> **Day 3:** The egg hatches and absorbs its *chorion* (outer shell), and the newly hatched larva lays down on the base of the cell and is fed royal jelly exclusively.

>> **Days 4–8:** The cell containing the developing larva remains open, and the larva is fed royal jelly by nurse bees. The cell is extended downward and elongated into a vertical shape, sometimes described as looking like a peanut shell. (Larvae destined to become worker bees remain in cells parallel to the ground, and they are fed a different food that causes them to develop into a worker bee.) ***Note:*** Day 4 is the best time for a young larva to be selected to become a queen rather than a worker.

>> **Day 9:** The queen cell is capped, and the developing queen (pupa) consumes royal jelly that the workers stored in the cell with the larva. She spins a cocoon inside the cell.

>> **Days 10–14:** The developing queen transforms into its adult form. Her body is soft and *very* fragile.

>> **Day 15:** The developing queen is less fragile. ***Note:*** At this point the queen cell can be carefully moved to a queenless nuc.

>> **Day 16:** The queen emerges from her cell. ***Note:*** Cool temperatures can slow the queen's development, as can extreme heat. Under such conditions, a queen could take 17 or 18 days to emerge.

**TIP**

A virgin queen takes a few days to mature — her wings expand and dry, her glands mature, and so on. Then she needs a few days more to fly and mate, and a few days more to settle down to laying eggs. Allow two or three weeks from emergence to the time when she will begin laying eggs.

# Talking about the Birds and Bees for Honey Bees

It's important to understand how honey bees mate so you can do your best to provide optimal conditions and know how circumstances such as weather can impact your queen-rearing operation. The queen bee has some interesting mating habits:

>> **Queen bees mate in the air.** The drones fly out of their colonies and gather at a place called a *drone congregation area*. The virgin queens also know where these places are and make a "bee-line" there to mate. In this drone congregation area, several drones mate with the virgin. The drones then drop dead afterward (see Chapter 2 for more).

>> **A virgin queen will take one or more mating flights over the course of a few days or a week.** Then she's done mating for her lifetime. The sperm (from the drones) is stored in a special tiny ball in the queen's abdomen called the *spermatheca*. It is supplied with nutrients to keep the sperm alive for as long as the queen remains productive.

>> **Because queens mate with a number of drones, a honey-bee colony is a collection of "sub-families."** All the bees in the colony have the same mother (the queen); but some workers will be full sisters (having the same mother and father), and some will be half-sisters (having the same mother but different fathers). This genetic diversity is critical to having thriving, healthy colonies with a variety of traits that help the bees survive.

>> **If for some reason a virgin is prevented from mating, there will come a time when she will stop trying to mate and will begin laying eggs.** However, none of these eggs will be fertilized, so they will all result in drones.

A queen doesn't mate with drones from her own colony. Doing so would result in undesirable inbreeding. The bees have developed a system to encourage genetic diversity and reduce the possibility of inbreeding. Instinctively, the queen flies much farther to drone congregation areas than do drones. So the queen goes farther from her hive to mate than the drones from her hive (her brothers) are willing to travel. This dramatically reduces the probability of her mating with a drone from her own hive.

# Creating Demand: Making a Queenless Nuc

A *nuc*, or *nucleus* colony, is a small community of bees. Most common are nucs of four or five frames. Chapter 4 includes information on nuc hives. A *queenless nuc* is a small colony of bees without a queen. If a colony doesn't have a queen, the bees will try to make a queen from any available larvae of the right age (four days after the egg was laid). A queenless nuc will also be receptive to a newly introduced virgin or mated queen.

Queenless nucs are used for starting queen cells and also for receiving queens — virgin or mated. Queenless nucs should be made up a day or two before they are needed so the bees will have time to realize that they no longer have a queen.

Here's how to create a queenless nuc:

1. **Place frames of capped brood, honey, and pollen into the nuc hive body. The frames should be covered with worker bees and come from a healthy hive.**

   **WARNING**

   Make 100 percent certain there is no queen on these frames — just workers (an occasional drone is okay). There should be as little open (uncapped) brood as possible.

   **TIP**

   Bees found on frames of open brood (not capped) are typically nurse bees. You can ensure that your queenless nuc has lots of nurse bees by shaking or brushing these nurse bees off of open brood frames into your queenless nuc. Just make sure you don't shake off or brush a queen into the nuc! Return the brood frame to the colony it came from.

2. **Wait a day or two before introducing a frame of eggs, a frame of grafted larvae, or a queen (virgin or mated).**

# Queen-Rearing Method 1: Go with the Flow

One good way to raise queens is to simply go with the bees' natural inclinations. Beekeepers may try to avoid swarming (Chapter 10 has information on swarming), and oftentimes the measures work. But swarming is something bees have done for eons. It's been a successful survival strategy for them, and try as you might to avoid it, you will at some point find swarm cells in your colony.

These cells will either be capped or open. Capped queen cells have been fed all they're going to get, and the queen inside is pupating. If they're still open at the bottom and contain an egg or a larva and a quantity or royal jelly, that's a cell on its way to producing a queen.

So imagine you inspect your colony, and you find some queen cells. This is a nice, healthy colony, so you decide to turn this into a queen-rearing opportunity. Rather than destroy the cells in an attempt to prevent swarming, fetch a trusty nuc box (or two) and make a split (or two).

## If the queen cells are capped

A swarm typically departs when the replacement queen cells are capped. So finding some or all of the queen cells capped means the swarm — and your old queen — has likely departed or soon will. Removing all the queen cells in this situation could render the colony unable to re-queen itself.

Instead of going on a mission to seek and destroy all those queen cells, you can capitalize on them instead. You need an extra hive setup for this, which can be a nuc box or a full-size hive.

Remove a frame that has queen cells on it. Add a frame or two of bees and brood, a frame or two of honey/pollen, and maybe a frame or two of foundation. If you're putting this nuc into a full-size hive, upsize the ratios to fill the box.

If you have multiple frames with queen cells, and lots of bees — and enough equipment — you can split off more than one nuc. If the queen cells are capped, meaning the original queen will soon go or is already gone, be sure to leave a frame with queen cells in the original (parent) hive so it, too, can continue to re-queen itself.

## If the queen cells are open

If you find queen cells that are all still open at the bottom, it's possible that the original queen is still in the house. It's best to find her so you're certain she's still around. You may want to move her to a nuc, along with some frames of bees, brood, nectar/pollen — and *no queen cells.* Let her serve as backup queen while you see how the other unit(s) play out. If there are queen cells on more than one frame, you can make another split into another nuc box with a frame containing queen cells, plus some other frames of bees, brood, food, and foundation. Just be sure to leave some queen cells in the original hive so those bees can raise a queen for themselves.

## Mind the timeline

From the time the queen cells are first capped, it's a week until the virgin(s) emerge. A virgin queen takes a few days — up to a week — inside the hive to mature. Her wings dry and harden. Then she takes her mating flights for one or more days, and if it rains or is too cold to fly, she passes on the mating flight and waits for better weather, which could take a few days or up to a week. Finally, after she's completed her mating flights, she settles down and begins to lay eggs. Her progress may be slow at first, but again, in a few days or a week's time, she'll have laid a big enough patch of eggs on one or more frames that you'll be able to see them. So from the time the queen cells are capped, it can be three or maybe even four weeks until you see eggs. Be patient.

If, after four weeks, you don't see any eggs, it's probably time to intervene. You can introduce a queen obtained elsewhere, or you can combine the queenless unit with another, *queen-right unit* (a colony that has a laying queen).

REMEMBER

A good reason to start more than one of these nucs is that if one of the queens doesn't work out, you can simply combine those bees back into another nuc unit. Use the newspaper method to do this, as explained in Chapter 18.

If, after four weeks, you have eggs in every split you made, you can celebrate with a glass of mead! You can grow each of these into full-size hives. If there's enough time left in the current season, you can add more nuc colonies that summer. If it's midsummer or later, they can be wintered as smaller, single hive-body colonies and then given more space the following spring. If this has brought you to a place where you have more colonies than you want, you can offer one or more nuc colonies to another beekeeper. Or you can sell one or more of the queens and combine the remaining bees back into the full-size hives you keep.

# Queen-Rearing Method 2: The Miller Method

The Miller Method is a queen-rearing process that requires no special equipment and is perfect for the backyard beekeeper who just wants to raise a few queens.

Here's how it works:

1. **First, take a deep frame with wax foundation and cut the bottom edge of the foundation into a sawtooth pattern (see Figure 14-2).** If the wax has wires in it, you'll need to snip a few of them or work around them.

**TIP**

The size of frame you use is not that critical. Most queen-rearing beekeepers use the same size frame used in their brood box. For most folks this would be a deep frame, although some beekeepers standardize on medium-depth frames for both brood and honey.

**FIGURE 14-2:**
Cutting a sawtooth pattern along the bottom edge of the wax foundation.

*Courtesy of Howland Blackiston*

2. **Place the frame with the sawtooth foundation in the center of your queen mother colony (the strongest, hardiest, most productive, and gentlest).** Let the bees draw it out into comb. Consider feeding the colony some sugar syrup to get them making wax. The syrup (like a nectar flow) stimulates wax production.

3. **After a week, have a peek every few days. At some point, the queen will start laying eggs in this new comb.** When the cells along the saw-toothed margins have eggs, it's time to set up a queenless nuc that will build and raise the queen cells.

4. **The day after setting up the queenless nuc, insert the sawtooth frame of eggs into the center.** Overnight, the bees will have become aware that they have no queen. They will be ever-so-ready to receive a frame containing just what they need to raise some queens: eggs and very young larva. If all goes well, the bees will build several queen cells along the jagged edge.

5. **In a week's time, have a look and see what the bees have made.** Hopefully, the bees will have built several queen cells in different spots along the jagged edge.

    In a few more days, come back and see how the queen cells are developing. Later, you will separate the capped queen cells by cutting them away from the comb. But for now, just look and see what you have to work with. If some of the cells are too close to cut apart, then plan to leave them together and put that clump in a queenless nuc. If the bees have managed to raise any queen cells on other frames, you should destroy them. They're not from your carefully chosen queen mother, and you want the bees' attention focused on raising the daughters of that favored queen.

6. **Make up an additional queenless nuc for each queen cell (or clumps of cells) that were present in Step 5.**

7. **A few days before the queens are due to emerge, go back and remove the frame containing the queen cells.**

   Cut the cells apart carefully to put into the waiting queenless nucs (see Figure 14-3). When you cut the comb, take plenty of comb around each cell or clump of cells — give yourself a generous handle, even if it means cutting into other brood cells. Don't dent or deform the queen cell in the least little bit — the developing queen inside is extremely fragile. Also, don't tip the cells or jostle them for the same reason.

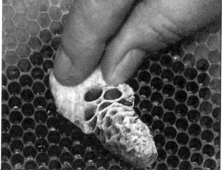

**FIGURE 14-3:** Carefully cut around the queen cell and transfer it onto a frame in a waiting queenless nuc (left). By gently pressing on the extra handle, it should stick securely in place (right).

*Courtesy of Howland Blackiston*

**WARNING**

Be sure to move those queen cells to the queenless nucs before the queens emerge. If you don't, the first queen to emerge will kill all the other queens and queen cells in the hive! A sad ending after so much effort.

8. **Distribute the queen cells to the queenless nucs.**

   Remove a central frame from each queenless nuc and carefully press the comb handle attached to the cells into the comb of this frame. The cells should be hanging vertically. Be very careful as you slide the frame back into the hive. Not denting the queen cell is of paramount importance.

9.  **A day or two after "emergence day" (16 days after the egg was laid), check to see that each queen did emerge from her cell.**

    You'll see the queen cell with a round opening on the bottom. You may be able to find her walking around on one of the frames. Then again, you may not. A virgin queen is often on the small side, not much larger than a worker (she'll plump up after she mates). If you don't find her, don't worry. A virgin emerging from a cell into a queenless colony is very likely to be accepted — hey, she's the only game in town.

# Queen-Rearing Method 3: The Doolittle Method, also Known as Grafting

With the Miller Method you can raise a few queens at a time, but other methods allow you to ramp up production and turn out dozens of high-quality queens. Raising larger numbers of queens is more challenging, but you might find a market for your lovely queens among beekeepers in your area. Whatever the market is charging for queens, you can charge that also. Feel free to charge more if you think you have some hot property. Right now, queens sell for around $30 each. I've seen truly superior queens selling for much more.

The most common method of producing large numbers of queens is by grafting larvae of the right age into special wax or plastic *queen cell cups* that are affixed to bars. The bars are positioned in frames, and the frames are inserted into a queenless nuc equipped with lots of nurse bees and lots of provisions such as honey (and/or syrup) and pollen (and/or pollen substitute).

## Tools and equipment

Grafting requires some special equipment and supplies. All are available from most beekeeping suppliers. If you're a gadget lover, grafting has a lot of allure.

>> **Cell bar frames:** These are frames that, instead of containing foundation, contain one or more bars that hold plastic or wax queen cups into which larvae are grafted. The frame is then inserted into a queenless colony where queen cells will be raised. See Figure 14-4.

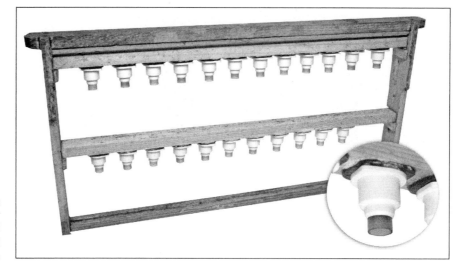

**FIGURE 14-4:**
A cell bar frame (detail of individual queen cup).

*Courtesy of Howland Blackiston*

>> **Grafting tools:** You use grafting tools (see Figure 14-5) to lift the delicate and oh-so-fragile larva out of its original cell and place it gently in the cup on the cell bar frame. There are many different kinds of grafting tools, and each beekeeper develops his or her own preference, often making personal modifications.

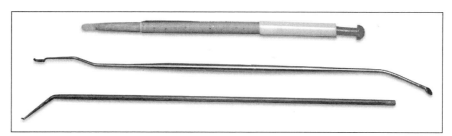

**FIGURE 14-5:**
Three different kinds of grafting tools.

*Courtesy of Howland Blackiston*

>> **Queen cell protectors:** Cell protectors are cage-like cylinders that are placed around the developing queen cells once they are capped (see Figure 14-6). They look kind of like hair curlers. The cell protectors keep the newly emerged virgin queens confined, preventing them from moving about the colony and killing the other queens. The perforations in the protector allow surrounding worker bees to feed and care for the new virgin queens.

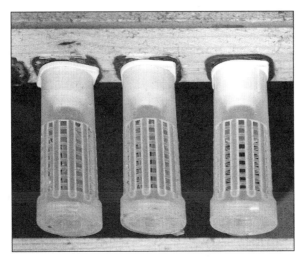

**FIGURE 14-6:**
Queen cell
protectors
snapped into
place. They are
used to protect
queen cells
before queen
emergence and
to hold new
queens that
emerge from
their cells.

*Courtesy of Howland Blackiston*

>> **Queen cages:** These are designed to confine the queen and provide, via screen or perforations, a way for the bees outside of the cage to feed the queen inside (see Figure 14-7). Queen cages also have a place that can be filled with a fondant or candy plug, which will be eaten by the bees and allow the queen to be released after sufficient time has passed to have her accepted by the colony.

**FIGURE 14-7:**
Different kinds of
cages designed to
hold her majesty.

*Courtesy of Howland Blackiston*

# How it's done

There are steps leading up to and following grafting day. As with the Miller Method, you select the colony headed by your best queen for grafting.

>> **Four days before grafting day:** The eggs you'll want to graft are laid four days before grafting day. To make it easier to locate the right-age larva, confine the queen on a frame of empty drawn comb four days ahead of grafting day. Use that comb when transferring larvae to cell cups. A *push-in* cage is the perfect tool for restricting a laying queen to a cluster of empty cells. See Figure 14-8.

**FIGURE 14-8:** A push-in queen cage helps you confine a queen to just a few cells. The eggs laid in these cells are the ones you want to use for grafting.

*Courtesy of Howland Blackiston*

>> **Three days before grafting day:** Release the queen from confinement by removing the push-in cage. Having laid eggs in these cells, the queen's job is done. She can be allowed to roam the colony and continue laying eggs at will. You will want to keep track of where you placed the queen cage because those larvae are the ones that will be the right age for grafting. Mark the frame's top bar so you can retrieve that frame come grafting day. I make a written notation on the top bar using a permanent marker.

» **Two days before grafting day:** Create your queenless nuc to serve as a cell starter. You want to put your freshly grafted larvae into an environment where they'll be well cared for. This means lots of bees (especially lots of nurse bees), frames of honey, pollen (and/or a feeder and a pollen patty), and little or no open brood. You want lots of nurse bees because they are the ones most geared to feeding larvae.

» **Grafting day:** Using the frame you confined the queen on four days ago, graft larvae into cell cups and place the frame of cells into the queenless cell starter that you made a couple of days ago. Grafting is a delicate maneuver, and the very young larvae are exceedingly fragile. See Figure 14-9.

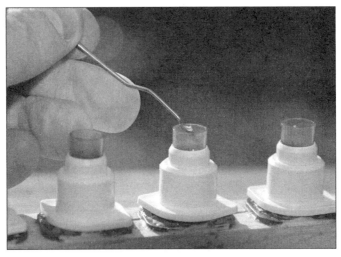

*Courtesy of Howland Blackiston*

**WARNING**

The larva breathes through small openings on one side of its body — the side exposed to the air. If the larva is flipped when transferred to the cell cup, it will not be able to breathe.

**TIP**

When you're familiar with what larva look like when they're grafting age, you can skip the steps of confining the queen; just look through the queen mother's hive and select a frame with larvae of the appropriate age.

» **One or two days after grafting day:** Have a peek. The bees have decided which cells they're going to feed and draw and develop into queens and which cells they are not. You will quickly notice the difference.

>> **A week or so after grafting day:** Check the cells, and when they are capped, put cell protectors on them. The cells should be capped by the bees four or five days after grafting, and the only care they need from that point until emergence is warmth and humidity. The cell protectors will keep the virgin queens separated from each other. If they are left to emerge into the colony, the first queen out will kill all the other queens.

**WARNING**

During your inspection a week following grafting day, give the other frames a look and remove any rogue queen cells elsewhere that the bees may have built in the hive. If one of them emerges, she'll kill all the other queens.

Virgin queens will emerge 15 to 17 days after the egg is laid (11 to 13 days after grafting). The average development time is 16 days, but development is faster in warmer weather and slower in cooler temperatures, so there is some variation in timing.

**TIP**

To achieve successful grafting, be prepared to try various grafting tools, different positions (for you and for the frame of larvae), different lighting, and perhaps some magnification. Try a few different tools, and give yourself time and a lot of practice to get it right. A larva that is bruised or handled too much is unlikely to survive and be accepted by the bees. You may have poor acceptance your first few rounds of grafting, but keep trying. Eventually you'll get the knack.

## Providing nuptial housing

Virgin queens are often given temporary housing until they mate. Mating nucs for your virgins can be regular-sized hives or nucs, but if you're raising a lot of queens, that can be quite demanding in terms of bees, equipment, and real estate. Small units, called *mini-mating nucs* or *pee wees,* hold just a couple of cups of bees, a few miniature frames, and a small food reservoir (see Figure 14-10). With mini-mating nucs, you can place a bunch of nucs in a comparatively small area. Checking these smaller units is quicker and easier, too. But be careful — in mid-summer when nectar falls off, it can be hard to keep these tiny colonies from being robbed.

After the queens have emerged safely into their cell protectors, you can transfer them to queen cages (with candy plugs). Then introduce the caged queens into their own mating nuc (one queen per nuc), letting the bees slow-release the queens by chewing through the candy plug.

Be patient. Allow two to three weeks for a virgin queen to mate and begin laying eggs.

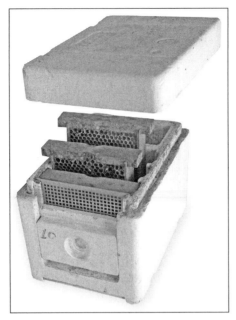

*Courtesy of Howland Blackiston*

# Finding Homes for Your Queens

Now that you have queens, what do you do with them? Where do they live? You have several options:

>> **Queen cages:** Your queens can be kept for a few days in a cage with a few attendants. Over a week is too much stress on the queen, and she may die. Use a clean eyedropper to feed them a single drop of water once or twice a day.

>> **Battery box:** This is a special ventilated container (usually used for shipping queens). It can also be used to house your new queens for a few days. The battery boxes typically hold 20 queen cages and have a space for a slab of sugar fondant (food) and a sponge (water). Be certain to add a cupful or two of attendant bees (a hundred or more) who will care for the queens.

>> **Queen banks:** A *queen bank* is a regular, queen-right colony with a queen excluder keeping the colony's resident queen separate from the upper hive body. In the upper hive body, caged queens are held in a special rack that takes the place of one of the frames. Attendants from the colony care for the caged queens. Using this method, queens can be housed for several days or even a couple of weeks (versus a few days, as with the other two options).

When queens are in the bank, help ensure that they're surrounded by nurse bees by periodically (weekly) moving frames of young, open brood from the lower portion up to the bank portion. Examine the frames carefully as you move them up — be careful not to move the queen from the lower portion up to the queen bank portion! If nectar is scarce, feed the queen bank with sugar syrup using whatever kind of feeder you prefer.

You can store mated queens for days or a few weeks in a queen bank. However, the sooner they get out and start laying eggs, the better off they'll be. They're meant to be roaming comb among a population of daughter worker bees and laying eggs. Confinement is hard on them, so get them out into a colony as soon as possible.

# Evaluating the Results

So you went to all the trouble to raise queens that you thought would be as good as or better than where you started. How did it turn out?

Whenever you work with bees, it's good to make a habit of constantly evaluating the quality of the colony. For bees in your queen-rearing operation (large or small), formalize these observations by keeping some sort of record. Write it down! This record can take the form of a notebook or a set of index cards — whatever works for you. Even writing on the outer cover can be helpful, though there is a limit to how much information will fit there. It's a good idea to keep track of what queen was this queen's mother. Over time, you may want to purposely switch from one queen mother to an unrelated (but still superb) queen to avoid inbreeding. Evaluate temperament, monitor Varroa mite buildup, and make note of any other signs of disease or distress. Notice how they build up in the spring and handle a nectar dearth. Your notes can be simple — a series of check, check-plus, check-minus — or you can develop a more detailed analysis. There's a lot of information to be managed, and having the information written down will help immeasurably. Don't trust your memory!

You can review these notes to decide which queens to favor in your next round of queen rearing and which to eliminate. You can compare various queens in terms of the traits you selected for. In this way, you can fine-tune your technique and be sure you move toward continuous improvement.

# The Queen Rearer's Calendar

**Spring:** This is swarming season and an ideal time to raise queens. The bees are very willing to raise and accept new queens. Nectar and pollen are plentiful. Keep an eye on your drone mother colonies.

When drone cells are emerging, start actively rearing queens. The drones will be mature and ready for mating by the time your virgins are out on their mating flights.

**Summer:** Keep on raising queens. Be aware that as nectar flow diminishes, the colonies will need feeding. In late summer especially, feeding is necessary not only to keep the smaller nucleus colonies afloat, but to aid in queen acceptance.

**Fall:** Queen rearing is done. Time to get nucs settled and ready for winter. Consider using a side-by-side nuc that can house two queens.

**Winter:** Now's a good time to read your notes and make plans for the coming season.

You can set up some nucleus colonies for your new queens and overwinter them by setting them on top of other full-size colonies. This is especially helpful in cooler, northern climates. Heat rises, and the full-size colonies will help keep the nucs warm. The following spring, these overwintered nucs will have young queens that can be used to repopulate colonies that die over the winter. Or you can grow these nucs into full-sized colonies.

Overwintering nucs does require some special equipment: a split bottom board and a hive body divider of some kind (a division board and/or a division feeder).

# Marking Your Queens

It's a good idea to mark your mated queens. Beekeeping suppliers often sell queen-marking pens, or you can buy similar pens at an art supply store (just be careful to get water-based, permanent paint pens). An international queen-marking color code has been established to track the age of a given queen:

| For years ending with | Mark with this color |
|---|---|
| 0 or 5 | Blue |
| 1 or 6 | White |
| 2 or 7 | Yellow |
| 3 or 8 | Red |
| 4 or 9 | Green |

# 5 Sweet Rewards

Find out how honey is made, where it originates, what it consists of, and why it's so good for you.

Discover the six basic types of honey and myriad varietals that give honey its unique colors and flavors.

Learn how to taste and evaluate honey for culinary appreciation and competitive judging.

Decide what type of honey you want to produce, how and when to harvest and package it, and how to go about marketing it.

Find out about the special equipment you will need on honey harvest day, how to use it, and how to care for it between seasons.

Chapter **15**

# Honey, I Love You

*S*pecial thanks to C. Marina Marchese for her work in preparing this chapter. Marina is co-author (with Kim Flottum) of The Honey Connoisseur *and is founder of the* American Honey Tasting Society.

These days honey has a huge following. You can hardly pick up a magazine without finding some reference to this remarkable product of the honey bee. It's used in beauty care, sought for its nutritional benefits, revered for its medicinal properties, valued as an ingredient for culinary applications, and expertly paired with foods by top chefs around the world. Honey's uses are fantastically diverse. It's not just for biscuits and tea anymore!

## Appreciating the History of Honey

Man has used honey for thousands of years. Certainly as a sweetener, but more often as a medicine to treat a variety of ailments. It has an illustrious history finding its way into folklore, religion, and every culture around the world. In ancient times, honey was considered a luxury good enjoyed by the privileged and royals, as well as a valid form of currency to pay taxes.

The Cuevas de la Araña (Spider Caves) of Valencia, Spain, are a popular tourist destination. These caves were used by prehistoric people who painted images on the stone walls of activities that were a critical part of their everyday life, such as goat hunting, and lo and behold, honey harvesting. One of the paintings depicts two individuals climbing up vines and collecting honey from a wild beehive. Immortalized on the rock wall between 6,000 and 8,000 years ago, it is widely regarded as the earliest recorded depiction of honey gathering. See Chapter 1 for an image of this cave drawing.

Ancient Egyptians were the first known nomadic beekeepers to migrate their beehives on boats up and down the Nile in order to follow the seasonal bloom specifically for pollination (see Figure 15-1). Beeswax paintings on the pyramid walls depicted beekeepers smoking their hives and removing honey. This tells us that the Egyptians understood the seasonality of beekeeping and the symbiotic relationship between honey bees and pollination.

In 2007 a remarkable find was made during an archaeological dig in the Beth Shean Valley of Israel. An entire apiary was uncovered from Biblical times, containing more than 30 mostly intact clay hives, some still containing very, very old bee carcasses. These man-made hives date from the 10th to early 9th centuries BCE, making them the oldest man-made beehives in the world. It is believed that as much as a half-ton of honey was harvested each year from this site.

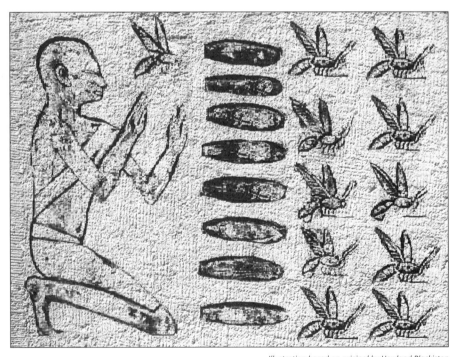

**FIGURE 15-1:** Egyptians harvested honey over 3,000 years ago.

*Illustration based on original by Howland Blackiston*

# Understanding the Composition of Honey

Honey is the sweet result of the bees magically transforming the nectar they gather from flowers. Honey is about 80 percent fructose and glucose, and between 17 and 18 percent water. Maintaining a balance between sugar and water is critical to the quality of honey. Excess water, for example from poor storage, can trigger the spontaneous yeast to ferment, and the honey will spoil. The bees nail this balance instinctually, but our improper harvesting and storing of honey can upset the delicate ratio.

More than 20 other sugars have been found in honey. There are also proteins in the form of enzymes, amino acids, minerals, trace elements, and waxes. The most important enzyme is *invertase* — which is an enzyme added by the bees. This is responsible for converting the sugar sucrose (found in nectar) into the main sugars found in honey: fructose and glucose. It is also instrumental in the ripening of the nectar into honey.

With an average pH of 3.9, honey is relatively acidic, but its sweetness hides the acidity.

The antibacterial qualities associated with honey come from hydrogen peroxide, which is a by-product of another enzyme (glucose oxidase) introduced by the bees. See "Healing with Honey" later in this chapter.

The plants themselves and the soil they grow in contribute the minerals and trace elements found in honey. See Figure 15-2 for a typical breakdown of honey content.

## Composition*

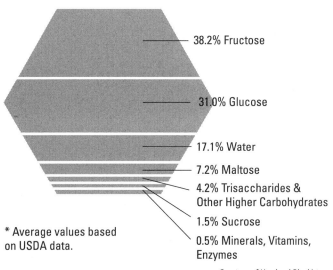

38.2% Fructose

31.0% Glucose

17.1% Water

7.2% Maltose

4.2% Trisaccharides & Other Higher Carbohydrates

1.5% Sucrose

0.5% Minerals, Vitamins, Enzymes

* Average values based on USDA data.

**FIGURE 15-2:** This chart illustrates the typical content of honey (based on data from the USDA).

*Courtesy of Howland Blackiston*

Honey owes its delicate aromas and flavors to the various volatile substances (similar to essential oils) that originate from the flower. As heat decomposes the fructose, hydroxymethylfurfural (HMF) naturally found in all honeys increases, thus lowering the quality. Each of these components that make up honey is extremely fragile, and overheating honey or improper or long-term storage can compromise not only the healthful benefits but also its flavors (see "Raw versus regular honey" later in this chapter).

# Healing with Honey

Honey is nice on toast. But did you know it will also relieve a multitude of medical issues — from taming a cough and alleviating allergies to healing cuts or burns? Because of its low pH and hygroscopic properties, bacteria cannot survive in honey. The pollen in the honey contains various minerals as well as enzymes and B vitamins, which impart immune-boosting properties that help the body fight infection. Good stuff.

TIP

Generally speaking, the darker the honey, the greater the minerals and antibacterial qualities.

If you are sensitive to sugars, the fructose and glucose in honey aid in maintaining blood-sugar levels. The first gives you a natural burst of energy, and the second sustains your blood levels so you won't get the sugar blues as you might with processed white sugar. Overall, honey is a wise choice — it is wholesome, flavorful, and the only naturally processed sweetener found in nature.

## Honey and diabetes

All kinds of conflicting information can be found on the Internet about whether honey is or is not okay for those with diabetes. My advice is that if you are on a diabetes eating plan, you should check with your doctor for the definitive answer. In the meantime, the following information from the Mayo Clinic's website is helpful:

> Generally, there's no advantage to substituting honey for sugar in a diabetes eating plan. Both honey and sugar will affect your blood-sugar level. Honey is sweeter than granulated sugar, so you might use a smaller amount of honey for sugar in some recipes. But honey actually has slightly more carbohydrates and more calories per teaspoon than does granulated sugar — so any calories and carbohydrates you save will be minimal. If you prefer the taste of honey, go ahead and use it — but only in moderation. Be sure to count the carbohydrates in honey as part of your diabetes eating plan.

## Honey's nutritional value

One tablespoon of honey (21grams) provides 64 calories. Honey tastes sweeter to most people than sugar, largely due to the higher fructose level, and as a result, most people likely use less honey than they would sugar.

Honey is also a rich source of carbohydrates, providing 17 grams per tablespoon, which makes it ideal for your working muscles because carbohydrates are the primary fuel the body uses for energy. Carbohydrates are necessary to help maintain muscle glycogen, also known as stored carbohydrates, which are the most important fuel source for athletes to help them keep going.

## Honey and children

WARNING

The medical community advises that children should be at least 18 months before introducing honey into their diets. Spores of botulism naturally find their way into any raw agricultural product or simply the dust that may settle inside a honey jar. Mature digestive and immune systems can normally handle this type of bacteria; however, children and infants under 18 months of age should not consume raw honey. Every individual responds uniquely to ingesting honey, so seek advice from a qualified medical care provider.

# Choosing Extracted, Comb, Chunk, or Whipped Honey

What style of honey do you plan to harvest? You have several different options (see Figure 15-3). Each affects what kind of honey harvesting equipment you purchase, because specific types of honey can be collected only by using specific tools and honey-gathering equipment. If you have more than one hive, you can designate each hive to produce a different style of honey. Now that sounds like fun!

TIP

Honey should never be refrigerated because cold temperatures accelerate crystallization. In time, however, nearly all honeys form granulated crystals, regardless of the temperature. There is nothing wrong with honey that has crystalized. It's perfectly okay to eat in this form. However, if you prefer, crystallized honey can be easily liquefied by placing the jar in warm water or by gently heating it in the microwave on a low setting for a minute or two.

**FIGURE 15-3:** The four basic types of honey you can produce (extracted, comb, chunk, and whipped).

## Extracted honey

Extracted honey is by far the most popular style of honey consumed in the United States. Wax cappings are sliced off the honeycomb, and liquid honey is removed (extracted) from the cells in a honey extractor by centrifugal force. The honey is strained and then put in containers. The beekeeper needs an uncapping knife, extractor (spinner), and some kind of sieve to strain out the bits of wax and the occasional sticky bee. Chapter 17 has instructions on how to harvest extracted honey.

## Comb honey

Comb honey is honey just as the bees made it . . . still in the beeswax comb. In many countries, honey still in the comb is considered the only authentic honey untouched by humans while retaining all the pollen, propolis, and health benefits associated with unadulterated honey straight from the hive. When you uncap those tiny beeswax cells, the honey inside is now exposed to the air for the very first time since the bees stored it. Encouraging bees to make this kind of honey is a bit tricky. You need a strong nectar flow to get the bees to make comb honey. Watch for many warm, sunny days and just the right amount of rain to produce a bounty of flowering plants. But harvesting comb honey is less time-consuming than harvesting extracted honey. You simply remove the entire honeycomb and package it. You eat the whole thing: the wax and honey. It's all edible! A number of nifty products facilitate the production of comb honey (but more on that in Chapter 16).

TIP

Top Bar hives are great for producing beautiful comb honey. Find out more about Top Bar hives in Chapter 4.

## Chunk honey

Chunk honey refers to placing slabs of natural honeycomb in a wide-mouthed bottle and then topping the jar off with extracted liquid honey. Packing chunks of honeycomb into a glass jar is a stunning sight, creating a stained-glass effect when a light-colored honey is used. By offering two styles of honey in a single jar (comb and extracted), you get the best of two worlds.

## Whipped honey

Also called *creamed honey, spun honey, churned honey, candied honey,* or *honey fondant,* whipped honey is a semisolid style of honey that's popular in Europe and Canada. In time, all honey naturally forms coarse granules or crystals. But by controlling the crystallization process, you can produce very fine crystals and create a velvety-smooth, spreadable product.

*Granulated honey* is honey that has formed glucose crystals. You make *whipped* honey by blending nine parts of extracted liquid honey with one part of finely granulated (crystallized) honey. Store in a cool place and stir every few days. The resulting consistency of whipped honey is thick, ultra-smooth, and can be spread on toast like butter. Making it takes a fair amount of work, but it's worth it!

Most honeys naturally crystallize — except tupelo, black sage, acacia, and honeydew honey. When honey crystallizes, the texture can be smooth as silk or somewhere in between coarse sand and a cat's tongue. This natural occurrence begins when the glucose spontaneously loses water and forms a solid crystal around pieces of pollen that have settled at the bottom of the jar. As more crystals grow, the honey slowly turns solid. If the texture of your honey is not pleasing to you, you can restore it to a liquid state by gently warming the entire jar by placing it in hot water.

### Honey or honeydew honey?

There is another style that does not originate from floral nectar. It's known as *honeydew honey.* It's commonly found in the EU and is revered in religious text under the name *manna.* When various other insects (typically aphids) gather nectar from flowers, leaves, and tree buds, honey bees will gather these insects' secretions. The resulting honeydew honey made by bees is rich in minerals and enzymes. It is generally dark in color and prized for its many health and healing benefits around the world. It's sometimes difficult to tell the difference between traditional honey and honeydew simply by tasting it. Often honeydew sold in shops is labeled as *oak, forest,* or *pine* honey. It is *not* prone to crystallization because of its high pH.

# Taking the Terror out of Terroir

The floral sources, soil, and climate ultimately determine how your honey will look, smell, and taste. Winemakers call these ever-changing environmental factors *terroir.* The sensory characteristics of any honey — color, aroma, and flavor profile — are reflective of a specific flower from a specific region. Location, location, location! This unexpected diversity of each honey harvest makes tasting honey a never-ending culinary experience. Opportunities to visit honey producers and taste the honey of that region and season can be an eye-opening and delicious adventure.

The flavor of honey your bees make is likely more up to the bees than you. You certainly can't tell them which flowers to visit. See Chapter 3 for a discussion of where to locate your hive when you want a particular flavor of honey.

## Customizing your honey

REMEMBER

Unless you put your hives on a farm with acres of specific flowering plants, your bees will collect myriad nectars from many different flowers, which results in a delicious honey that's a blend of the many flowers in your area. This type of honey is classified as *wildflower honey.* Note also that eating such honey is an effective way to fend off local pollen allergies — a natural way of inoculating yourself. Wildflower honey is what most of us will produce as backyard beekeepers.

Some beekeepers harvest honey from a single floral source, resulting in *varietals* or *uni-floral* honeys. Granted, several acres of that single floral source are needed, and the bees must be prepared to work the bloom at the very moment it is producing nectar. But what we get from this focused approach are honeys that have distinctive flavor profiles that resemble the flower and region from which the honey is harvested. Simply put, varietal honeys are poetry in nature (see Figure 15-4).

**FIGURE 15-4:** Some of the honey varietals from around the world that are in honey connoisseur Marina Marchese's collection.

*Courtesy of Marina Marchese*

# Honey from around the world

Here is a short list of some of the most popular varietal honeys harvested by beekeepers around the world. I include some notes on color, aroma, and flavor as well as my favorite food pairings for these various honeys.

**Acacia**
**Region:** Europe, Asia, Africa
**Color:** Transparent pale-yellow color
**Aroma:** Delicate
**Flavor:** Sweet notes of apricot and pineapples
**Pairings:** Salty cheeses, fresh grapefruit, or grilled salmon
**Notes:** Acacia blooms in early spring.

**Alfalfa**
**Region:** Midwest
**Color:** Light amber, straw

**Aroma:** Dry hay, sour, pungent
**Flavor:** Barnyard, spicy, yeasty
**Pairings:** Cornbread, grilled lamb chops
**Notes:** Grown for hay to feed livestock, alfalfa apparently doesn't want honey bees to steal its nectar. So it's evolved the ability to "spank" the bee when she tries to enter the nectary. This disciplinary action deters the bees and prompts them to chew holes into the side of the flowers to get to the nectar.

### Avocado
**Region:** Mexico, California, Texas, Florida, and Hawaii
**Color:** Dark amber
**Aroma:** Medium intense, smoky, brown sugar
**Flavor:** Fruity, burnt sugar, nutty
**Pairings:** Barbecue

### Blueberry Blossom
**Region:** Maine, Michigan, and Wisconsin
**Color:** Light to medium amber
**Aroma:** Warm milk, musty fruit
**Flavor:** Dark fruit, notes of blueberries
**Pairings:** Soft cheeses, ice cream, or pound cakes
**Notes:** Blueberry pollination is the second-largest annual pollination event in the United States. (Almond pollination is the largest.)

### Buckwheat
**Region:** The largest producers are Russia and China, although states in the northeastern region of the United States also farm buckwheat, as well as Illinois, Maryland, Michigan, Minnesota, North Carolina, Ohio, Oregon, South Dakota, West Virginia, and Wisconsin.
**Color:** Dark amber
**Aroma:** Woody, musty, malty
**Flavor:** Chocolate, coffee, dark beer
**Pairings:** Barbecue, dark cherries

### Chestnut
**Region:** Italy, Spain, France, and around Europe
**Color:** Dark amber
**Aroma:** Woody, pungent, intense
**Flavor:** Bitter, vegetal, nutty
**Pairings:** Poached pears, walnuts, or panna cotta
**Notes:** Blooms appear in June until July, yet the nut matures later in the season. Festivals are common to celebrate the dark earthy and sometimes bitter chestnut honey harvest.

**Clover**
**Region:** Heartland USA
**Color:** Light amber, straw
**Aroma:** Green grass, beeswax, spicy
**Flavor:** Cinnamon, grassy, dry hay
**Pairings:** Salad dressings, rice pudding, or honey butter
**Notes:** Clover honey can be found in just about every grocer and kitchen because it's the premier floral source for honey bees in the United States.

**Fireweed**
**Region:** Pacific Northwest and Alaska
**Color:** Transparent; crystallizes slowly
**Aroma:** Delicate, dried fruit, brown sugar
**Flavor:** Pears, warm caramel, and pineapples
**Pairings:** Fried chicken, carrots, gingersnaps
**Notes:** Fireweed is known as a colonizer plant because it's the first plant to grow after a wildfire. Without other plants nearby, it is easy to harvest pure fireweed honey. Fireweed blooms in June and lasts until September.

**Lavender**
**Region:** The Provence area of France
**Color:** Light straw
**Aroma:** Delicate, floral notes
**Flavor:** Almond, vanilla
**Pairings:** Nougats
**Notes:** The vast fields blanket the south of France, where everything is inspired by the scent of lavender.

**Linden** or **Basswood**
**Region:** United States, United Kingdom, and Europe, known as lime tree honey
**Color:** Medium golden amber
**Aroma:** Butterscotch, fruity, herbal
**Flavor:** Green fruit, butterscotch with a signature menthol finish
**Pairings:** Green melon and fresh mint, seared scallops, and cilantro
**Notes:** Many city streets are lined with linden trees offering city beekeepers a delicious harvest. Basswood trees are good nectar producers.

**Ling Heather**
**Region:** The moors of Ireland and Scotland
**Color:** Dark amber, red tint
**Aroma:** Intense, smoky, woody
**Flavor:** Toffee, bitter coffee, plum
**Pairings:** Steel-cut oats, cheddar cheese, or spice cake

**Notes:** Ling honey is so thick that it must be hand-pressed from the honeycomb. When inside the jar, Ling honey becomes thick as jam until stirred to become liquid again. This thick-to-thin property is called *thixotrophy*. Only a few rare honeys exhibit this quality.

### Locust
**Region:** East central United States, Europe, China
**Color:** Light-colored honey that has a green hue
**Aroma:** Delicate, vanilla
**Flavor:** Butterscotch, warm vanilla, and dried fruit flavors
**Pairings:** Salty cheeses
**Notes:** Blooms in early spring, showing off its highly aromatic clusters of white flowers. This honey is difficult to obtain because of its early bloom time. Conditions must be perfect.

### Manuka
**Region:** New Zealand and southeast Australia
**Color:** Beige; granulates quickly
**Aroma:** Musty, camphor, earthy
**Flavor:** Earthy, damp, and evergreen — like an Alpine forest
**Pairings:** Used internally for ulcers and health reasons
**Notes:** Honey is world renowned for its healing properties (especially regarding ulcers and intestinal problems). Manuka is the name for the tea tree, which grows in New Zealand and southeast Australia. Manuka is mostly taken as a medicine here in the United States. It is thixotropic and has a very high viscosity. *Warning:* Because of the very high premium paid for manuka, much of the commercially available product labeled as manuka is not what it claims to be. New Zealand produces 1,700 tons of manuka honey annually (nearly all the world's production comes from New Zealand). But 10,000 tons or more of honey labeled as "manuka" is being sold annually. That doesn't quite add up!

### Mesquite
**Region:** Texas, Louisiana, Arizona, New Mexico, and along the Mexican border
**Color:** Medium amber
**Aroma:** Warm brown sugar
**Flavor:** Smoky, woody
**Pairings:** Barbecue, pecan pie

### Orange Blossom
**Region:** Tropical regions, predominately in Southern California and Florida
**Color:** Burnt orange
**Aroma:** Floral notes of jasmine and honeysuckle
**Flavor:** Orange, perfumy flowers
**Pairings:** Cranberry bread, glazed chicken, or red cabbage slaw

**Notes:** The flavor of orange honey harvested in the West has a warmer flavor reminiscent of the sandy desert.

**Sage**
**Region:** Native to the United States and can be found along the dry, rocky coastline and hills of California
**Color:** Water white
**Aroma:** Delicate, sweet
**Flavor:** Delicate, sweet notes of green and camphor
**Pairings:** Roasted potatoes, lamb, or watermelon and feta
**Notes:** A distinct taste of the warm western desert

**Sidr**
**Region:** Hadramaut in the southwestern Arabian peninsula
**Color:** Dark amber
**Aroma:** Cooked fruit
**Flavor:** Rich, dates, molasses, green fruit
**Pairings:** Traditional warm breads, pastries
**Notes:** Considered the finest and most expensive honey in the world (expect to pay around $300/pound). It contains the highest amount of antioxidants, minerals, and vitamins. Beekeepers laboriously carry their hives up the mountains twice each year. It's the Rolls Royce of honey!

**Sourwood**
**Region:** Called "mountain honey" because the tree is located throughout the Blue Ridge mountain region
**Color:** Water white, gray tint
**Aroma:** Warm, spicy
**Flavor:** Nutmeg, cloves ending with a sour note
**Pairings:** Figs, blue cheese, or warm cider
**Notes:** A highly sought-after honey

**Star Thistle**
**Region:** Found across the entire United States
**Color:** Yellow amber with a green tint
**Aroma:** Cooked tropical fruits, wet grass
**Flavor:** Green bananas, anise, and spicy cinnamon
**Pairings:** Manchego cheese, vinaigrette, herbal teas
**Notes:** Attractive plant for honey bees; often called spotted knapweed

**Thyme**
**Region:** Highly praised in Greece
**Color:** Very light amber
**Aroma:** Pungent

**Flavor:** Fruit, resin, camphor, herbs
**Pairings:** Pastries and sweet cakes
**Notes:** It is also known as Hymettus honey, named after the mountain range near Athens.

**Tupelo**
**Region:** Grows in the swamps of southern Georgia and northern Florida
**Color:** Light golden amber, it rarely crystallizes
**Aroma:** Spicy, buttery
**Flavor:** A delightful mix of earth, warm spice, and flowers
**Pairings:** Buttermilk biscuits, honey-mustard salmon
**Notes:** Often referred to as the champagne of honey, you'll pay champagne prices for this rare variety.

# The Commercialization of Honey

On average, the United States consumes 450 million pounds of honey a year. And yet American beekeepers produce only around 149 million pounds a year. Hmm. Where does that honey come from to meet the demand? It comes from overseas. Countries like Argentina, China, Germany, Mexico, Brazil, Hungary, India, and Canada export millions of pounds of honey each year to satisfy America's sweet tooth.

## Is it the real deal?

Unbeknownst to the average consumer, honey is one of the top foods targeted for adulteration and fraud. Unfortunately, to fill the huge demand, some commercial honey producers and importers do unscrupulous things like cutting honey by adding extraneous, improper, or inferior ingredients. They may add high fructose corn syrup in order to extend their assets. Some remove the naturally present pollen in their honey by heating and ultra-filtering it. This makes it sparkling clear and less prone to crystallization (see the following section for the difference between *raw* and *regular* honey). Filtering also makes the source of the honey difficult to trace, because the pollen allows the honey to be tracked back to the floral source and the region where it was produced.

## Raw versus regular honey

The main difference between *regular* and *raw* honey is that commercially produced honey (such as that found in supermarkets) is typically pasteurized and ultra-filtered. Pasteurization is the process where honey is heated at high temperatures to kill any yeast that may be present that may cause botulism. It's also done to keep the honey from crystallizing, making it look more attractive to consumers. In addition, the ultra-filtering process removes pollen (and makes the product sparkling clear). But all this heating and filtering destroys most of the enzymes and some vitamins and removes the beneficial pollen. It also evaporates the natural aromas and flavors. So regular commercially pasteurized honey doesn't have as many health benefits or sensory pleasures as raw honey. Raw may not look as commercially attractive, but raw honey has more flavor and aroma than its pasteurized counterpart. It also means the wonderful health benefits are not compromised.

## Organic or not?

Some honey is labeled as *organic*. To put such a claim on a label in the United States, the producer must be certified as organic. It's a great marketing idea (after all, organic products are very marketable). But the claim of being organic is not necessarily an accurate representation. Part of the problem is that the United States Department of Agriculture (USDA) has not yet developed definitive guidelines for what constitutes organic honey, but it is working to amend the current and somewhat vague USDA regulations concerning the production of organic beekeeping products (for example, honey). This new action will establish USDA certification standards specifically for managing honey-bee colonies and bee products (currently the criteria is for livestock, generally). The scope of this new action includes specific provisions for transition to organic apiculture production, replacement of bees, hive construction, forage areas, supplemental feeding, healthcare practices, pest-control practices, and an organic apiculture system plan.

Given that bees forage nectar and pollen at will from flowers that are miles away from the hive, there is no practical way to guarantee these flowering plants are not being subjected to chemical treatments or that the plants are not genetically modified. It will be interesting to see what the USDA comes up with. For now, I suggest you take any organic claims with a grain of salt.

For the latest status of the new organic beekeeping regulations (known as Organic Apiculture Practice Standard, NOP-12-0063), visit the US General Services Administration (GSA), Office of Information and Regulatory Affairs, at www.reginfo.gov.

## Your own honey is the best

The commercialization of honey is one more reason why keeping honey bees and producing your own honey is the sweeter choice. You know how the product was produced, how you care for your bees, and from where the bees gathered their nectar. Alternatively, purchase honey from local farmers' markets or wherever small-scale regional beekeepers are selling their product. Ask your beekeeper about his bees and management practices. He'll be happy to give you a taste before you buy a jar.

# Appreciating the Culinary Side of Honey

When I first tasted fresh honey from the hive, I was immediately impressed with how mild and subtle the flavor was and was surprised that honey could simply be harvested and eaten without any other special preparations. This single, uncomplicated fact made the whole idea of keeping honey bees deliciously appealing to me.

I like to tell people that tasting honey is like tasting wine, except you don't have to spit. The tasting technique for honey is comparable to tasting wine, coffee, tea, or chocolate. You use all of your senses to learn more about what you experience in your mouth. You simply take note of each honey's color, aroma, clarity, texture, taste, and (most important) flavor notes.

## The nose knows

Humans experience thousands of flavors with our noses; however, we are only capable of experiencing four basic taste sensations on our tongues — sweet, sour, salty, and bitter. Of our five senses, our sense of smell is approximately a thousand times more sensitive than our sense of taste; everything else is considered flavor. Try this exercise: Hold your nose closed as you put a spoonful of honey on your tongue. What do you taste? A sweet liquid but no real flavor? Now, unplug your nose and inhale; immediately you smell its flavor. When you open your nose, the aromas go up your nose into your olfactory bulb, which is where you taste.

Next, pour a few tablespoons of honey into a small glass jar, cup both hands around it to warm the honey, and with a spoon swirl it around the edges to move the molecules. Notice the honey's color and texture. There are seven designated colors of honey — water white, extra white, white, extra light amber, light amber, amber, and dark amber. Now stick your nose inside the jar and take a deep smell. This is the best way to capture the honey's aromas. Take a spoonful onto your tongue and let it melt; then inhale to smell the honey's flavors. Can you identify the flavors? You can use words like floral, fruity, grassy, or woody. Figure 15-5 shows the seemingly unlimited range of aromas and tastes you can experience when you're sampling honey. The more honey you taste, the more you will determine its multitude of flavors. Taste, aroma, and texture experienced together are the main components that impart flavor in your mouth.

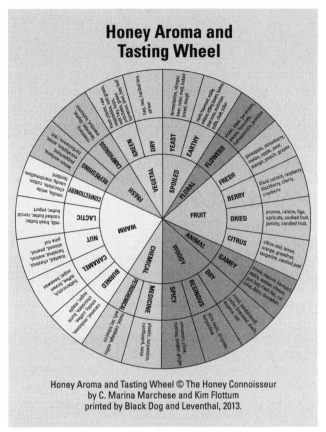

**FIGURE 15-5:**
Use this honey aroma and tasting wheel to pinpoint the unique flavors and characteristics of your honey and others.

*Excerpted with permission from The Honey Connoisseur, by C. Marina Marchese and Kim Flottum. Black Dog & Leventhal Publishers, 2013.*

## Practice makes perfect

No one is born an expert taster; learning to recognize flavors comes from the practice of conscious tasting. The repetitive act of tasting different samples of honey side by side is where the discovery comes into play, and you quickly begin to recognize the differences and similarities between each honey. Whereas sugar and other sweeteners are simply sweet, honey can express floral, grassy, fruity, or woody flavor notes. Look for a wide range of flavors that are clearly identifiable. Take note on when the flavors appear while the honey is on your tongue — there's a beginning, middle, and finish to each flavor.

## Recognizing defects in honey

On occasion, you may come across a honey with undesirable flavors that are not considered positive attributes. These are called off-flavors, or defects. Some are easily recognized, and others require a bit of experience to identify. Common defects are fermentation from a high percentage of water present or improper storage; smoke or burnt flavors from old wax comb or too much smoke used at harvest; or metallic tastes from rusty equipment or storage in metal containers. There also can be chemical residues from colony treatments or dirty equipment and extracting techniques. Any of these are offensive and unwelcome. If it tastes bad, something is wrong. Always keep in mind that pure honey has layers of flavor, regardless of its type and flavor profile.

# Pairing Honey with Food

There is never a wrong way to eat honey. It pairs perfectly with every food group, and sometimes it is best enjoyed simply off the spoon. You will find that some food pairings will quickly become your favorites. Honey served with cheese is a timeless classic. This favorite pairing can be traced back to a Roman gourmand named Marcus Gavius Apicius (first century AD).

Begin with foods that have flavors and textures you enjoy. Try fresh pears, figs, or walnuts with bread or crackers. Now choose a few varietal honeys or your own harvest, and drizzle over the pairing. Look for combinations that complement or contrast with the honey. Sometimes they blend in your mouth to create an entirely new tasting experience. When one overpowers the other or cancels out another flavor, you have a clash in your mouth. A creamy goat cheese complements a buttery and fruity honey. A rich, dark honey contrasts nicely with a stinky bleu cheese. Serve honey with bread and crackers and sides like fresh or dried fruits, nuts, and vegetables to add color and texture. The choices are endless, and you'll have fun serving up your favorites at your next gathering.

For some outstanding recipes using honey, be sure to buzz over to Chapter 20.

# Infusing Honey with Flavors

Honey will take on the flavors of any foods that you infuse into the jar. It's a delicious experiment to add flavors like citrus zest, ginger, rosemary, cinnamon, or even rose petals to your honey to add an extra dimension. Be sure to wash and completely dry any food you infuse into your honey. The drying prevents added water from upsetting the delicate water and sugar percentage, which can cause the honey to ferment. Let your infusion sit at room temperature for two to three weeks — the longer it sits, the more flavorful your infused honey will become.

**TIP**

I like to fill a glass jar with shelled, lightly salted walnuts and top it off with extracted honey. Now that makes for one seriously good ice-cream topping! And a welcome hostess gift to present to friends.

# Judging Honey

Preparing honey for competition is an exciting step for beekeepers to demonstrate their attention to detail at the show bench. Honey judges are trained to scrutinize each entry for a perfect presentation of honey samples. Judges follow strict guidelines that are understood by the entrants. Air bubbles that create unsightly foam during extraction, debris, wax, lint from cheesecloth during straining, overly dark comb from propolis, early crystallization, fermentation, and even small holes in the wax cappings are just a few errors that can disqualify even the most delicious honey. Judges look for defects and take note of all these factors on score cards that serve to aid the entrant in understanding how and why the judges came to their final decision. Moisture is a significant factor in fermentation of samples and can be quickly determined by a hand-held refractometer (see Figure 15-6).

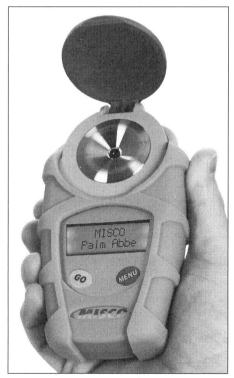

**FIGURE 15-6:** A refractometer is used to measure the water content of honey.

*Courtesy of Misco Refractometers*

Surprisingly, identifying flavors is not as highly important to the judging process as a seamlessly prepared sample with a pleasing flavor. Look for honey competitions in your local region. You'll also find national and even worldwide honey shows. The rules, regulations, and judging criteria vary from show to show, so be certain to ask for a copy of the judging criteria and a honey application.

# Honey Trivia

Here are some tidbits of "betcha-didn't-know" information about honey that will make for some good banter at the dinner table:

>> How many flowers must honey bees tap to make one pound of honey? *Two million.*

>> How far does a hive of foraging bees fly to bring you one pound of honey? *Over 55,000 miles.*

>> How many flowers does a honey bee visit during one nectar collection trip? *50 to 100.*

>> How much honey would it take to fuel a bee's flight around the world? *About one ounce.*

>> How long have bees been producing honey from flowering plants? *10 to 20 million years.*

>> What Scottish liqueur is made with honey? *Drambuie.*

>> What is the U.S. per capita consumption of honey? *On average, each person consumes about 1.3 pounds per year.*

Interesting, huh? How about three more little-known honey facts:

>> In olden days, a common practice was for newlyweds to drink mead (honey wine) for one month (one phase of the moon) to ensure the birth of a child. Thus the term *honeymoon*.

>> One gallon of honey (3.79 liters) typically weighs 11 pounds, 13.2 ounces (5.36 kilograms.).

>> Honey found in the tombs of the Egyptian pharaohs was still edible when discovered centuries later. That's an impressive shelf life!

# Chapter **16**

# Getting Ready for the Golden Harvest

t all comes down to honey. That's why most people keep bees. For eons honey has been highly regarded as a valuable commodity. And why not? No purer food exists in the world. It's easily digestible, a powerful source of energy, and simply delicious. In many countries, honey even is used for its medicinal properties. The honey bee is the only insect that manufactures a food we eat. And we eat a lot of it — more than 1 million tons are consumed worldwide each year.

What a thrill it is to bottle your first harvest! You'll swear that you've never had honey that tastes as good as your own. And you're probably right. Commercial honey can't compare to homegrown. Most supermarket honey has been cooked and ultrafiltered; these processes make the product more visually attractive for commercial sale. And alas, some commercial honey isn't even pure honey, having been blended with less-expensive sweeteners to increase profits. Yours will be just the way the bees made it — pure, natural, and packed with aroma and flavor. I'm getting hungry just thinking about it.

TIP

There is so much to say about honey, and Chapter 15 is devoted exclusively to this topic.

In this chapter, I help you plan for the big day — your first honey harvest. You need to consider the type of honey you want to produce, the tools you need, the amount of preparation to do, and what you need for marketing. So, time to get started.

# Having Realistic Expectations

In your first year, don't expect too much of a honey harvest from your Langstroth hive. Sorry, but a newly established colony doesn't have the benefit of a full season of foraging. Nor has it had an opportunity to build its maximum population. I know that's disappointing news. But be patient. Next year will be a bonanza!

Beekeeping is like farming. The actual yield depends on the weather. Many warm, sunny days with ample rain result in more flowers and greater nectar flows. When gardens flourish, so do bees. If Mother Nature works in your favor, a hive can produce 60 and even up to 100 pounds of surplus honey (that's the honey you can take from the bees). If you live in a warm climate (like Florida or Southern California), you can expect multiple harvests each year. But remember that your bees need you to leave some honey for their own use. In cold climates, leave them 60 pounds; in climates with no winter, leave 20 to 30 pounds.

TIP

If you're planning to harvest from a Top Bar hive, be aware that these hives are simply not geared for large-scale honey production. That's not to say that you can't get a small harvest to share with family and friends year after year. The good news is that the harvesting using a Top Bar hive is all low-tech, and there's no heavy lifting. At the end of this chapter, I include some Top Bar harvesting techniques.

# What Flavor Do You Want?

The flavor of honey your bees make is likely more up to the bees than you. You certainly can't tell them which flowers to visit. See Chapter 3 for a discussion of where to locate your hive when you want a particular flavor of honey, and see Chapter 15 for a description of the characteristics and flavors of more than 20 domestic and international varietals of honey.

REMEMBER

Unless you put your hives on a farm with acres of specific flowering plants, your bees will collect myriad nectars from many different flowers, which results in a delicious honey that's a blend of the many flowers in your area. Your honey can be classified as *wildflower honey.* Note also that eating such honey is an effective way to fend off local pollen allergies — a natural way of inoculating yourself. See Chapter 15 for more information on different kinds of honey.

You need to decide what *style* of honey you want to package, because that influences some of the equipment that you use. The names of the different styles of honey include *extracted, comb, chunk,* or *whipped.* These are reviewed in detail in Chapter 15. Of these, the most popular styles are extracted and comb, so this chapter focuses on these two in particular. Extracted is the easiest of these two for the new backyard beekeeper, and that's where I suggest you start. Instant gratification! After you get the first season under your belt, by all means try your hand at comb honey and the other styles.

# Assembling the Right Equipment to Extract Honey

When you decide what style of honey you want your bees to make, you need to get the appropriate kind of equipment. This section discusses the various types that you need to extract honey.

## Honey extractors

I suggest that those with Langstroth hives start with extracted honey. For this, you need an extractor, a device that spins honey from the comb using centrifugal force (see Figure 16-1). Extractors come in different sizes and styles to meet virtually every need and budget. Hand-crank models or ones with electric motors are available. Small ones for the hobbyist with a few hives, huge ones for the bee baron with many hives, and everything in between can be found. Budget extractors are made entirely of plastic, while rugged ones are fabricated from food-grade stainless steel. Keep in mind, however, that a good-quality, stainless-steel extractor will far outlast a cheap one made of plastic. So get the best one that your budget allows. Look for a model that accommodates at least four frames at a time. Backyard beekeepers can expect to pay hundreds of dollars for a new, quality extractor. A used one costs less. Add on more dollars for ones with electric motors.

**TIP**

You may not have to buy an extractor. Some local beekeepers, beekeeping clubs, and nature centers rent out extractors. So be sure to call around and see what options you have. Ultimately, you may want to invest in your own. My advice: If you can, rent or borrow an extractor during your first season. From the experience you gain, you'll be better able to choose the model and style of extractor that best meets your needs.

## Uncapping knife

The wax cappings on the honeycomb form an airtight seal on the cells containing honey — like a lid on a jar. Before honey can be extracted, the "lids" must be removed. The easiest way is by using an electrically heated uncapping knife. These heated knives slice quickly and cleanly through the cappings (see Figure 16-2).

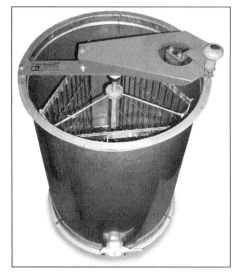

*Courtesy of Howland Blackiston*

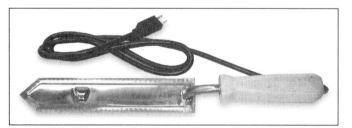

*Courtesy of Howland Blackiston*

Alternatively, you can use a less-costly serrated knife called a cold uncapping knife. Heat it by dipping in hot water (be sure to wipe the knife dry before you use it to prevent any water from getting into your honey).

## Honey strainer

The extracted honey needs to be strained before you bottle it. This step removes the little bits of wax, wood, and the occasional sticky bee. Any kind of conventional kitchen strainer or fine-sieved colander will suffice. Nice stainless-steel honey strainers (see Figure 16-3) are made just for this purpose and are available from your beekeeping supplier.

Or you can use a disposable cloth paint strainer (available at your local paint supply store). It does the trick just fine and fits nicely over a five-gallon, food-grade plastic bucket.

*Courtesy of Howland Blackiston*

# Other handy gadgets for extracting honey

Here are a few of the optional items that are available for extracting honey. None are essential, but all are useful niceties.

## Double uncapping tank

The double uncapping tank is a nifty device that is used to collect the wax cappings as you slice them off the comb. The upper tank captures the cappings (this wax eventually can be rendered into candles, furniture polish, cosmetics, and so on). The tank below is separated by a wire rack and collects the honey that slowly drips off the cappings. Some say the sweetest honey comes from the cappings! The model shown in Figure 16-4 also has a honey valve in the lower tank.

## Uncapping fork or roller

An uncapping fork (sometimes called a cappings scratcher) or an uncapping roller is used to scratch or punch open cappings on the honeycomb (see Figure 16-5). It can be used in place of or as a supplement to an uncapping knife (the fork opens stubborn cells missed by the knife or uncapping roller).

FIGURE 16-4:
A double
uncapping tank
helps you harvest
wax cappings. It
reclaims the
honey that drains
from the
cappings.

*Courtesy of Howland Blackiston*

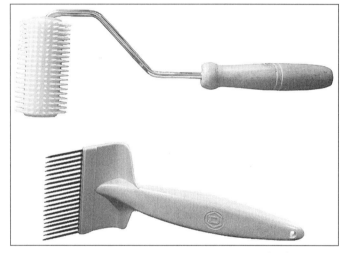

FIGURE 16-5:
An uncapping
fork or roller is a
useful tool for
opening cappings
missed by your
uncapping knife.

*Courtesy of Howland Blackiston*

## Bottling bucket

Five-gallon bottling buckets are made with food-grade plastic and include a honey gate at the bottom. They come with airtight lids and are handy for storing and bottling honey. Each pail holds nearly 60 pounds of honey. I always keep a few of them on hand (see Figure 16-6).

Courtesy of Howland Blackiston

**FIGURE 16-6:**
The honey gate
valve at the
bottom of this
five-gallon bucket
makes bottling
your honey a
breeze.

## Solar wax melter

Aside from honey, one of the most important products of the hive is beeswax. From the wax, you can make candles, furniture polish, and cosmetics. Your primary harvest of wax is the result of the cappings that you cut from the comb during the honey extraction process. These cappings (and any burr comb that you trim from the hive during the year) can be placed in a solar wax melter and melted. A single hive yields enough surplus wax to make a few candles and some other wax products.

Find recipes for making products from beeswax in Chapter 18.

TIP

You can obtain a *solar wax melter* by purchasing it from your bee supplier or by making one yourself. It typically consists of a wooden box containing a metal pan covered with a glass lid (see Figure 16-7). The sun melts the wax, which is collected in a tray at the base of the unit. It's a handy piece of equipment if you plan to make use of all that wax.

Find instructions for making your own solar wax melter in my book *Building Beehives For Dummies* (Wiley).

TIP

FIGURE 16-7:
You can use a solar wax melter to begin the rendering process. Very green!

*Courtesy of Brushy Mountain Bee Farm*

## Honey containers

Select an attractive package for your extracted honey (jar, bottle, and so on). Many options are available, and quite frankly, any kind of container will do. Clear containers are best because customers want to be able to see what they're getting. Either plastic or glass is okay to use. You can purchase all kinds of specialized honey bottles from your beekeeping supplier. Or simply use the old mayonnaise and jam jars that you've been hoarding. The key is to keep the jar air-tight. Honey is hygroscopic, meaning it attracts moisture from the air. Without an airtight seal, the additional water content will eventually cause the honey to ferment and spoil.

# Planning Your Extracted Honey Harvest Setup

Giving some thought to where you plan to extract and bottle your honey is important. You can use your basement, garage, tool shed, or even your kitchen. You don't need a big area. If you have only a few hives, harvesting is a one-person job. But be prepared — you'll likely get plenty of volunteers who want to help out. The kids in my neighborhood are eager to lend a hand in exchange for a taste of my *liquid candy*. The guidelines in the following list will help you choose the best location:

>> **The space you choose must be absolutely bee-tight.** That is to say, you don't want any bees getting into the space where you're working. The smell of all that honey will attract them, and the last thing you want is hundreds (or thousands) of ravenous bees flying all about.

» **Never, ever attempt to harvest your honey outdoors.** If you do, disaster is imminent! In short order you'll be engulfed by thousands of bees, drawn by the honey's sweet smell.

» **Set up everything in advance, and arrange your equipment in a way that complements the sequential order of the extraction process (see Figure 16-8).**

Uncapping tank

Honey extractor

Bottling setup

**FIGURE 16-8:**
Here's the setup, left to right, in my garage for extracting honey.

*Courtesy of Howland Blackiston*

» **Have a bucket of warm water — better yet, hot and cold running water — and a towel at the ready.** Life gets sticky when you're harvesting honey, and the water is a welcome means for rinsing off your hands and uncapping knife.

» **If you're using an electric uncapping knife, you'll need an electrical outlet.** But remember that water and electricity don't mix well, so be careful!

» **Place newspapers or a painter's drop cloth on the floor.** This little step saves time during cleanup. If your floor is washable, that really makes life easy!

# Gathering Comb Honey Equipment

Harvesting comb honey boils down to two basic equipment choices: using section comb cartridges or the cut-comb method. Either works fine. You need special equipment on your Langstroth hives to produce these special kinds of honey. See Chapter 15 for additional information about comb honey production.

## Section comb cartridges

Honeycomb kits for Langstroth hives consist of special supers containing wooden or plastic section comb cartridges. Each cartridge contains an ultrathin sheet of wax foundation. Using them enables the bees to store honey in the package that ultimately is used to market the honeycomb. My favorite kit — Ross Rounds — makes circular section comb in clear plastic containers. This is a product with enormous eye appeal!

**TIP**

You typically need a strong nectar flow to encourage the bees to make any kind of section comb honey. For this reason, making comb honey is more challenging than the more traditional extracted honey. You may want to defer this adventure for the second year of your backyard beekeeping.

## Cut comb

The cut-comb method can be used with either Langstroth or Top Bar hives. In the case of Langstroth hives, a special foundation that is ultrathin and unwired is used in the frames. After bees fill the combs with capped honey, the combs are cut from the frames or top bars. You can use a knife or a *comb honey cutter,* which looks like a square cookie cutter and makes the job easy and the resulting portion uniform and attractive.

# Branding and Selling Your Honey

Before you harvest your first crop of honey, you may want to give some advanced thought to the label you will put on it. You may even want to sell your honey. After all, a hundred or more bottles of honey may accommodate more toast than your family can eat! The following sections describe some ideas to help you think this through.

# Creating an attractive label

An attractive label can greatly enhance the appearance and salability of your honey. It also includes important information about the type of honey and who packages it (you!). Generic labels are available from your beekeeping supplier. Or you can make your own custom label. I easily reproduce my labels (see Figure 16-9) using my computer's printer and an appropriate size of blank, self-adhesive labels.

**FIGURE 16-9:**
Here's my honey label — simple and to the point.

*Courtesy of Howland Blackiston*

**TIP**

The guidelines and requirements for proper honey labeling have been changing in recent years. There are all kinds of new do's and don'ts that should be considered if you plan to sell your honey commercially. For the latest and official info, contact the National Honey Board, 390 Lashley St., Longmont, CO 80501-6045. Its web address is www.nhb.org.

You should include a few important bits of information on your label (particularly if you're planning to sell your honey). The following are suggestions for you to consider:

>> **State what the container contains.** The word *honey* must be visible on the label, and the name of a plant or blossom may be used if it's the primary floral source. Honey must be labeled with its common or usual name on the front of your package, such as "Honey," "Wildflower Honey," or "Clover Honey."

>> **Include your name and address** (as the producer). Consumers should be able to contact you.

>> **Report the net weight of your product** (excluding packaging), both in pounds/ounces and in metric weight (grams). This information should be included in the lower third of your front label panel in easy-to-read type, such as Net Wt. 16 oz. (454 g). When determining net weight, use the government conversion factor of 1 ounce (oz) = 28.3495 grams or 1 pound (lb) = 453.592 grams. Round after making the calculation, not before. Use no more than three digits after the decimal point on the package. You may round down the final weight to avoid overstating the contents. When rounding, use typical mathematical rounding rules.

I recommend including information about the nutritional value (not usually required by law if you are producing fewer than 100,000 units per year or producing honey as a small company of fewer than 100 employees). But I think including this information makes the product far more professional looking. For more on the proper wording and design of a nutritional information label, see Figure 16-10.

In addition, consider the use of descriptive words and phrases, such as:

>> **Natural:** The word *natural* on the label can add value to the product in the eyes of many consumers. The Food and Drug Administration has a specific position on natural: nothing artificial or synthetic has been included or added that consumers would not expect to be in honey.

>> **Kosher:** The Hebrew word *kosher* means proper or fit. Kosher food must meet all the requirements of kosher dietary laws. To produce a kosher product, there are two areas of concern: raw materials, and equipment and production. Whereas honey in the hive is intrinsically kosher, it still needs to be certified because of processing. For products to be acceptable as kosher, they have to be certified by a recognized rabbinical authority whose approval is shown through recognized symbols on the label.

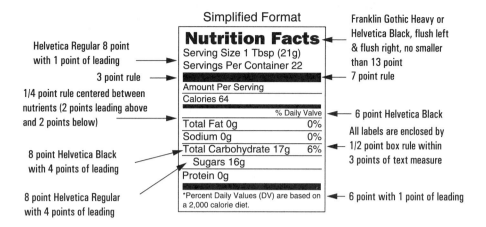

## Simplified Format

Helvetica Regular 8 point with 1 point of leading →

3 point rule →

1/4 point rule centered between nutrients (2 points leading above and 2 points below) →

8 point Helvetica Black with 4 points of leading →

8 point Helvetica Regular with 4 points of leading →

**Nutrition Facts**
Serving Size 1 Tbsp (21g)
Servings Per Container 22

Amount Per Serving
Calories 64

% Daily Value

| Total Fat 0g | 0% |
| Sodium 0g | 0% |
| Total Carbohydrate 17g | 6% |
| Sugars 16g | |
| Protein 0g | |

*Percent Daily Values (DV) are based on a 2,000 calorie diet.

← Franklin Gothic Heavy or Helvetica Black, flush left & flush right, no smaller than 13 point

← 7 point rule

← 6 point Helvetica Black

All labels are enclosed by 1/2 point box rule within 3 points of text measure

← 6 point with 1 point of leading

## Simplified Tabular Format

**Nutrition Facts**
Serving Size 1 Tbsp (21g)
Servings 22
Calories 64
*Percent Daily Values (DV) are based on a 2,000 calorie diet.

| Amount serving | %DV* | Amount serving | %DV* |
|---|---|---|---|
| Total Fat 0g | 0% | Total Carb. 17 g | 6% |
| Sodium 0 mg | 0% | Sugars 16 g | |
| | | Protein 16 g | |

## Linear Format

**Nutrition Facts** Serv size: 1 Tbsp (21g), Servings: 22, Amount Per Serving: **Calories** 64, **Total Fat** 0g (0% DV), **Sodium** 0mg (0% DV), **Total carb**, 17g (6% DV), **Sugars** 16g, **Protein** 0g, Percent Daily Values (DV) are based on a 2,000 calorie diet.

*Courtesy of Howland Blackiston*

**FIGURE 16-10:** These are the standard layouts and nutritional information that can appear on your label. Information in this example is based on a one-pound jar of honey.

>> **Honey and infants statement:** Although it's not required by law, many honey companies voluntarily add a statement on the honey label that says "Do not feed honey to infants under 18 months of age," or something similar. The reason for this is because honey may occasionally contain botulinum spores and has been identified as a risk factor for infant botulism.

**TIP**

Go to your local market and make mental notes about commercial honey labels. Which ones appeal to you? What about them makes them look so attractive? What kind of image or graphic is used? Which colors look best? Borrow ideas shamelessly from the ones you like best — but be careful not to steal anything that may be trademarked!

More detailed information about creating a distinctive label is available from the National Honey Board (see its contact info earlier in this section).

## Finding places to market your honey

An independently owned food market in your neighborhood may be interested in selling your honey. Honey is a pure and natural food, and you don't need a license to package and sell it as a backyard beekeeper (more detailed information is available from the National Honey Board; see the contact information in the previous section). Here are some other ideas:

>> Check out health food stores. They're always looking for a source of fresh, local honey.

>> Gift stores, craft shops, and boutiques are good places to sell local honey.

>> Put up an attractive sign in front of your house: HONEY FOR SALE (but check first to see if this is allowed in your community or if it might affect your insurance coverage).

>> Sell your honey at the local farmers' market.

>> Don't forget to consider church fairs, synagogue bazaars, and gardening centers.

>> And by all means *give* a bottle to all your immediate neighbors. It's the right thing to do and a great public relations gesture.

## Selling your honey on the web

This is the techie generation. So why not set up a website or use social media to promote and sell your honey all over the world? Remember that plastic honey jars are lighter to ship and less fragile than glass. This may be the time to invest in *Creating Web Pages For Dummies* by Bud E. Smith (Wiley) for more information on that particular subject. Now's the time to put on your best creative thinking cap and whip up an eye-catching site that will have visitors buzzing with excitement.

Chapter **17**

# Honey Harvest Day

The day you've anticipated all year is finally here, and it's time to reap the rewards of all your efforts (actually the bees did most of the work, but go ahead and take the credit anyway). It's time for the honey harvest!

And what better excuse could you ever have for having a party! Harvest day always is a big event at my house — a day when neighbors and friends gather to lend a hand and get a free sample from the magic candy machine. I schedule an open house on honey harvest day, inviting anyone interested to stop by, lend a hand, see how it's done, and go home with a bottle of honey still warm from the hive. I have plenty of honey-related refreshments on hand: honey gingersnaps, and honey-sweetened iced tea and lemonade.

I assume that you've decided to harvest extracted honey. That's what I recommend for new beekeepers of Langstroth hives. Harvesting extracted honey is easier for you and the bees, and you're far more likely to get a substantial crop than if you were trying to produce *comb honey.* Comb honey requires picture-perfect conditions to realize a successful harvest (large hive population, hard-working and productive bees, ideal weather conditions, and excellent nectar flows).

If you have a Top Bar hive, you will likely use the crush-and-strain method or a *honey press* to extract. If you have a Langstroth style of hive with removable wooden frames, you will use a spinning *extractor* to harvest your honey.

*Extracted honey* refers to honey that is removed from wax comb by either centrifugal force (using an extractor) or by crushing the comb and filtering out the honey (using a honey press). The honey is bottled in a liquid form, as opposed to harvesting comb or chunk honey where the honey isn't removed from the wax comb before it's packaged.

Be sure to allow yourself enough time for harvesting. I set aside an entire weekend for my harvest activities, part of one afternoon to get the honey supers off the hive and a good hunk of the following day to actually extract and bottle the honey.

# Knowing When to Harvest

Generally speaking, beekeepers harvest their honey at the conclusion of a substantial nectar flow and when the hive is filled with cured and *capped* honey (see Figure 17-1). Conditions and circumstances vary greatly across the country. Here in Connecticut, early one spring, I had an unusually large flow of nectar from a large honey locust tree. My bees filled their honey supers before June. I harvested this rare and delicate white honey in late May. I put the supers back on and got another harvest in the late summer. More typically, I wait until late summer to harvest my crop (usually mid-September). Where I live (in the northeastern United States), the last major nectar flow (from the asters) is over by September. First-year beekeepers are lucky if they get a small harvest of honey by late summer. That's because a new colony needs a full season to build up a large enough population to gather a surplus of honey.

*Surplus honey* refers to the honey that's beyond what the bees need for their own consumption. This extra amount of honey, stored in supers, is what you can harvest from the hive without creating any trouble for the colony. They made it just for you!

**TIP**

I suggest that you take a peek under the hive cover every couple of weeks during summer. Check in the honey supers that you placed on the hive earlier in the season. Note what kind of progress your bees are making and find out how many of the frames are filled with capped honey.

When a frame contains 80 percent or more of sealed, *capped* honey, you're welcome to remove and harvest this frame. Or, you can practice patience — leave your frames on and wait until one of the following is true:

>> The bees have filled all the frames with capped honey.

>> The last major nectar flow of the season is complete.

**FIGURE 17-1:**
On the top is a frame of comb from a Langstroth hive. On the bottom is honeycomb from a Top Bar hive. Note there is no wooden frame supporting this delicate comb on the sides and bottom, as there is with the Langstroth frame.

*Courtesy of Howland Blackiston with Jasmine and Zack Cecelic, www.wildhoodfarm.com*

**TIP**

Honey in *open* cells (not capped with wax) can be extracted *if* it is cured. To see if it's cured, and assuming you are using frames with foundation (versus Top Bar combs), turn the frame with the cells facing the ground. Give the frame a gentle shake. If honey leaks from the cells, it isn't cured and shouldn't be extracted. This stuff isn't even honey. It's nectar that hasn't been cured. The water content is too high for it to be considered honey. Attempting to bottle the nectar results in watery syrup that is likely to ferment and spoil. And just to be clear, if you have a Top Bar hive, this test is not possible. Top bar combs aren't sturdy enough to tolerate a shake — the comb may break off.

## Bad things come to those who wait!

You want to wait until the bees have gathered all the honey they can, so be patient. That's a virtue. However, don't leave the honey supers on the hive too long! I know, I know! Things tend to get busy around Labor Day. Besides spending a weekend harvesting your honey, you probably have plenty of other things to do.

But don't put off what must be done. If you wait too long, one of the following two undesirable situations can occur:

>> **After the last major nectar flow and when winter looms on the distant horizon, bees begin consuming the honey they've made.** If you leave supers on the hive long enough, the bees will eat much of the honey you'd hoped to harvest. Or they will start moving it to open cells in the lower deep hive bodies. Either way, you've lost the honey that should have been yours. Get those honey supers off the hive before that happens!

>> **If you wait *too* long to remove your supers, the weather turns too cold to harvest your honey.** In cool weather, honey thickens or even granulates, which makes it impossible to extract from the comb. I discuss this later in this chapter in the "Two common honey-extraction questions" sidebar. Remember that honey is easiest to harvest when it still holds the warmth of summer and can flow easily.

## A few pointers to keep in mind when harvesting liquid gold

Having enough honey jars and lids on hand is important. Standard honey jars are available in many different sizes and shapes. You can estimate that you'll harvest about 20 to 30 pounds of honey from each shallow honey super (assuming all the frames are mostly full of honey).

Honey is *hygroscopic* — meaning that it absorbs moisture from the air. On the positive side, this is why baked goods made with honey stay moist and fresh. On the negative side, this means you must keep your honey containers tightly sealed; otherwise, your honey will absorb moisture, become diluted, and eventually ferment.

TIP

Eventually all honey will crystalize. That does not change the purity of the product in any way. If you prefer liquid honey, you can heat the crystalized honey in a warm water bath (120 degrees Fahrenheit [49 degrees Celsius]) until it liquefies. Never heat honey excessively.

# Getting the Bees out of the Honey Supers

Regardless what kind of hive you have (frames or top bars), you must remove the bees from the honeycomb before you can extract or remove the honey. You've heard the old adage, "Too many cooks spoil the broth!" Well, you certainly don't need to bring several thousand bees into your kitchen!

**REMEMBER**

You must leave the bees 60 to 70 pounds of honey for their own use during winter months (less in those climates that don't experience cold winters). But anything they collect more than that is yours for the taking. That's the surplus honey I mention earlier.

**TIP**

To estimate how many pounds of honey are in your hive, figure that each deep frame of capped honey holds about 6 pounds of honey. If you have ten deep frames of capped honey, you have around 60 pounds! If you have a Top Bar hive (see Chapter 5 for an explanation of this and other hives), it's a little harder to estimate, because the actual length of the top bars and the depth of the comb can vary so much. Table 17-1 gives you a rough calculation for determining how much honey is in your hive.

**TABLE 17-1**

## Estimating Honey Yields

| Type of Honeycomb | Estimated Honey Capacity (Per Comb) |
| --- | --- |
| Langstroth deep frame | 6 pounds |
| Langstroth medium frame | 4 pounds |
| Langstroth shallow frame | 3 pounds |
| Warré Top Bar | 3 to 4 pounds* |
| Kenyan Top Bar | 4 to 6 pounds* |

*It's hard to be precise when estimating honey yield in Top Bar hives. The top bars can vary in length, and the free-form construction of the honeycomb can vary considerably.*

Removing bees from honeycomb can be accomplished in many different ways. This section discusses a few of the more popular methods that beekeepers employ. Before attempting any of these methods, be sure to smoke your bees the way you normally would when opening the hive for inspection. (See Chapter 7 for information on how to use your smoker properly.)

**WARNING**

If you are harvesting honey in the comb (producing comb honey), it's best not to smoke your bees. Bees tend to open the capped honey cells when smoke is applied. This can make an oozy mess of your pristine comb honey.

**WARNING**

The bees are protective of their honey during this season. Besides donning your veil, now's the time to wear your gloves. If you have somebody helping you, be sure he is also adequately protected.

# Shakin' 'em out

This bee-removal method involves removing frames (one by one) from honey supers and then shaking the bees off in front of the hive's entrance (see Figure 17-2). The cleared frames are put into an empty super. Be sure that you cover the super with a towel or board to prevent bees from robbing you of honey. You can use a bee brush (see Chapter 5) to gently coax any remaining bees off the frames.

**FIGURE 17-2:**
Shaking bees off a Langstroth frame.

**WARNING**

Don't try to shake bees off Top Bar comb. There's no wooden frame securing the wax comb (as there is when using a Langstroth frame). The unsupported comb on a top bar is very, very delicate. The shaking method will likely break the comb apart, so it's best to use the bee brush method to remove bees from the comb (see the "Getting the bees off Top Bar comb" section later in this chapter).

**TIP**

Note that the cells on comb tend to slant downward slightly — to better hold liquid nectar. Therefore, when brushing bees, you should always brush bees gently upward (never downward). This little tip helps prevent you from injuring or killing bees that are partly in a cell when you're brushing.

Shaking and brushing bees off frames aren't the best options for the new beekeeper because they can be quite time-consuming, particularly when you have a lot of supers to clear. Besides, the action can get pretty intense around the hive during this procedure. The bees are desperate to get back into those honey frames, and, because of their frenzy, you can become engulfed in a fury of bees. Don't worry — just continue to do your thing. The bees can't really hurt you, provided you're wearing protective gear.

# Blowin' 'em out

One fast way to remove bees from Langstroth supers (not recommended for Top Bar hives) is by blowing them out. Rest assured, they don't like it much. Honey supers are removed from the hive (bees and all) and stood on end. By using a special bee blower (or a conventional leaf blower), the bees are blasted from the frames at 200 miles an hour. Although it works, to be sure, the bees wind up disoriented and *very* irritated. Oh goodie. Again, I wouldn't recommend this method for the novice beekeeper.

TIP

A bee blower is basically the same as a conventional leaf blower, just packaged differently and usually more expensive.

# Using a bee escape board

Yet another bee-removal (and far less dramatic) method places method a bee escape board between the upper deep-hive body and the honey supers that you want to clear the bees from. Various models of escape boards are available, and all work on the same principle: The bees can travel down to the brood nest, but they can't immediately figure out how to travel back up into the honey supers. It's a one-way trip. (See Figure 17-3 for an example of a triangular bee escape with a maze that prevents the bees from finding their way back up into the honey supers.)

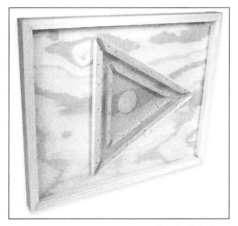

*Courtesy of Howland Blackiston*

TIP

Bee escapes work okay for Langstroth- and even Warré-style hives (see Chapter 5 for information on these and other hives), but it takes a couple days for the bees to be cleared from the honey supers. You must install the escape boards 48 hours before you plan to remove the honey supers. It takes that long to clear the bees. And, the thing is, you can't leave the escape board on for more than 48 hours, or the bees eventually solve the puzzle and find their way back into the supers. For me, a weekend beekeeper, the timing of all of this is quite impractical.

## Fume board and bee repellent

Here's my favorite method for a Langstroth hive — a fume board and bee repellent! It's fast and highly effective. And it's made even more desirable because of some wonderful products on the market (more about that later in this section).

A *fume board* looks like an outer cover with a flannel lining. A liquid bee repellent is applied to the flannel lining and the fume board is placed on top of the honey supers (in place of the inner and outer covers). Within five minutes, the bees are repelled out of the honey supers and down into the brood chamber. Instant success! The honey supers can then be safely removed and taken to your harvesting area.

TIP

This approach can be used for Langstroth- and Warré-style hives. It will not work with a Kenyan Top Bar hive.

In the past, chemicals used as repellents (either butyric propionic anhydride or benzaldehyde) have been hazardous in nature. They're toxic, combustible, and may cause respiratory damage, central nervous system depression, dermatitis, and liver damage. Need I say more? It's simply nasty stuff to have around the house.

In addition, the stench of each of these products is more than words can politely express. All of that has changed with products that effectively repel the bees but are nonhazardous and made entirely from natural ingredients (see Figure 17-4). Best of all, their scent smells good enough to be a dessert topping!

Here are step-by-step instructions for using a fume board with products such as Fischer's Bee-Quick or Bee-Dun:

1. **Smoke your Langstroth hive as you would for a normal inspection.**

2. **Remove the outer and inner covers and the queen excluder.**

3. **Use your smoker on the top honey super to drive the bees downward.**

4. **Spray the product on the fume board's felt pad in a zigzag pattern (as if corresponding to spaces between the frames) across the full width of the fume board.**

   Don't overdo it. About one ounce or less should do the trick (use more in cold, cloudy weather; less in hot, sunny weather). When in doubt, use less.

5. **Place the fume board on the uppermost honey super and wait three to five minutes for the bees to be driven out (this method works most effectively when the sun is shining on the fume board's metal cover).**

FIGURE 17-4:
A safe and fast
way to get bees
out of honey
supers is to use a
fume board with
products such as
Bee-Dun or
Fischer's
Bee-Quick.

*Courtesy of Howland Blackiston*

6. **Remove the fume board and confirm that most or all of the bees have left the super. They have? Good.**

7. **Now remove the top honey super. Put the super aside and cover it with a towel or extra hive cover to avoid robbing (discussed in the next section).**

8. **Repeat the process for subsequent honey supers.**

**TIP**

Keep in mind that a shallow super full of capped honey and frames can weigh 40 pounds. You'll have a heavy load to move from the beeyard to wherever you'll be extracting honey. So be sure to save your back and take a wheelbarrow or hand truck with you when removing honey supers from the hive. Figure 17-5 shows a hive carrier.

FIGURE 17-5:
If you have a
friend to help, a
hive carrier like
this makes
carrying heavy
supers and hive
bodies much
easier.

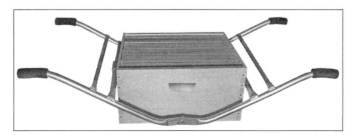

*Courtesy of Howland Blackiston*

# Honey Extraction from a Langstroth Frame

After the bees are out of the honey supers, prepare to process your honey as soon as possible (within a few days). Doing so minimizes the chance of a wax moth infestation, discussed in Chapter 13. Besides, extracting honey is easier to do when the honey is still warm from the hive because it flows much more freely.

REMEMBER

For a description of the various tools used in the honey-extraction process (uncapping knife, honey extractor or honey press, and so forth), see Chapter 16.

## Harvesting honey using an extractor

Follow this procedure when extracting honey from your Langstroth-style frames:

1. **One by one, remove each frame of capped honey from the super.**

   Hold the frame vertically over the double uncapping tank and tip it slightly forward. This helps the cappings fall away from the comb as you slice them.

2. **Use your electric uncapping knife to remove the wax cappings and expose the cells of honey.**

   A gentle side-to-side slicing motion works best, like slicing bread. Start a quarter of the way from the bottom of the comb, slicing upward (see Figure 17-6). Keep your fingers out of harm's way in the event the knife slips. Complete the job with a downward thrust of the knife to uncap the cells on the lower 25 percent of the frame.

Courtesy of Howland Blackiston

**FIGURE 17-6:**
Remove wax cappings using an electric uncapping knife. The uncapping tank below is a nifty accessory for collecting and draining the cappings.

3. **Use an uncapping fork (also called a *cappings scratcher*) to get any cells missed by the knife.**

   Flip the frame over, and use the same technique to do the opposite side.

TIP

   I discuss what you should do with the wax cappings, particularly if you want to use them for craft purposes, later in this chapter.

4. **When the frame is uncapped, place it vertically in your extractor (see Figure 17-7).**

   An *extractor* is a device that spins the honey from the cells and into a holding tank.

   After you've uncapped enough frames to fill your extractor, put the lid on and start cranking. Start spinning slowly at first, building some speed as you progress. Build speed gradually, without initially spinning the frames as fast as you can because extreme centrifugal force may damage the delicate wax comb. After spinning for five to six minutes, turn all the frames to expose the opposite sides to the outer wall of the extractor. After another five to six minutes of spinning, the comb will be empty. The frames can be returned to the shallow super.

**FIGURE 17-7:**
Place the
uncapped frame
vertically in the
extractor.

*Courtesy of Howland Blackiston*

5. **As the extractor fills with honey, it becomes increasingly difficult to turn the crank (the rising level of honey prevents the frames from spinning freely), so you need to drain off some of the harvest.**

   Open the valve at the bottom of the extractor and allow the honey to filter through a honey strainer and into your bottling bucket.

6. **Wait about an hour before bottling.**

   This resting period allows any remaining wax pieces and air bubbles to rise to the top.

7. **Use the valve in the bottling bucket to fill the jars you've designed for your honey.**

   Brand it with your label, and you're done! Time to clean up.

## Cleaning frames after extracting

Never store extracted frames while they're wet with honey. You'll wind up with moldy frames that have to be destroyed and replaced next year. You have to clean up the sticky residue on the extracted frames. How? Let the bees do it!

Here's a question I'm frequently asked: When I spin frames in my honey extractor, the entire unit wobbles uncontrollably, dancing across the floor. How do I prevent this from happening?

This shimmy happens when the load in the extractor is unbalanced. Make sure you have a frame in each of the basket's slots. Try redistributing the weight in the basket. Place frames of similar weight opposite each other. It's kind of like rearranging the wash load in the washing machine during the spin cycle. Some extractors can be bolted to a sturdy table — that helps.

Another frequently asked question I receive is: The honey in some of my shallow frames has granulated. I cannot remove the granulated honey using my extractor. How can I extract granulated honey from the honeycomb?

Unfortunately, no practical way to extract granulated honey from the honeycomb exists without destroying the honeycomb (melting the comb and wax, and then separating the wax once it floats to the top and solidifies). I suggest that you put supers with granulated honey back on the hives in the early spring. Before you do, scratch the cappings with an uncapping fork, exposing the honey to the bees. They'll consume the granulated honey and leave the combs sparkling clean and ready for a new harvest. **Note:** Don't feed your bees granulated honey in late autumn because it can give them dysentery.

At dusk, place the supers with the empty frames on top of your hive (sandwiched between the top deep and the inner and outer covers). Leave the supers on the hive for a few days and then remove them (you may have to coax any remaining bees from the supers by shaking them off the frames, or by using a bee escape or fume board). The bees will lick up every last drop of honey, making the frames bone dry and ready to store until next honey season.

# Harvesting Honey from Your Top Bar Hive

Before you can harvest honey from your Top Bar hive, you must determine whether your colony made excess honey and where it is in the hive.

**REMEMBER**

A Top Bar hive has honey stored in bands above the brood area on each comb. That honey is *not* part of the harvest and should be left for the bees' winter stores.

The honey you can harvest comes from those combs completely filled with honey and capped. Remember that these same capped honey stores can also be critical for the survival of the colony in winter. You must approach the harvest in a conservative way, always leaving enough to see the bees through the winter months.

## Selecting the comb to harvest

The pure honeycombs are likely to be found farthest from the entrance. In the case of a front-entrance colony, that would be toward the back.

You may notice that some of the honeycombs are capped on top, but the bottom still has some open cells. Those cells will have a higher moisture content and are not suitable to be included in the harvest. If a significant amount of your comb is uncapped, your honey is not ready and should be left in the colony until it's capped.

If just a few rows at the bottom of a comb are not capped, you can harvest those combs and discard the uncapped cells or, better yet, eat them on the spot.

**WARNING**

Since combs filled with honey are heavy, the bees will build extra bracing to the sidewall. This will be most apparent near the top just below the bar. This brace comb must be carefully detached before the comb is removed. As soon as you begin removing the brace comb, some honey will be released and bees will begin to lap it up. This is where planning and expedience come into play. Move quickly through the process and have a plan to cover the harvested combs immediately. You don't want to set off a robbing frenzy (see Chapter 10 for more on robbing frenzies).

Once you have identified what honeycombs you will take, the issue of handling the removed combs comes next. Top Bar combs filled with honey are very heavy and very fragile. The comb can break off the bar when mishandled. It is not recommended that you harvest on very hot days because the likelihood of breakage increases with higher temperatures. I also recommend that you build some kind of covered box or container that will allow you to transport the comb in a protected way to head off a robbing frenzy (see Figure 17-8).

## Getting the bees off Top Bar comb

The only way to remove bees from a Top Bar hive comb is to use a bee brush. Beekeepers use many objects for this purpose, including goose or turkey feathers, draftsmen's brushes, and the commercial brushes sold by bee supply houses. The best brushes can be rinsed off with water when they get sticky with honey.

FIGURE 17-8:
This box makes
it easy to
protect and
transport comb.

*Courtesy of William Hesbach*

Do bees like being brushed off comb? In a word, *no.* So it helps both you and the bees to brush them off using small, quick strokes with just enough force to remove the bees without damage to the bees or the comb. You want to direct them back into the hive body. The action can be thought of as a series of flicks as you work your way across the comb.

Both sides of the comb will have bees that need removing; remove the bees on the side facing you first and then lower the comb so you can reach over the back and remove the bees on the back. This technique is much easier than trying to turn the heavy comb with one hand.

## Harvesting using the crush-and-strain method

If you have just a few combs to harvest, your best option is to crush and strain them. Select a clean area and never harvest outside or any place that will allow the bees to find you. Line a strainer or colander with a paint strainer and place it over a clean bucket or a deep bowl.

The process is simple but decidedly messy. Slice or break a small section of comb off the top bar and squeeze it to release the honey. Use a clean pair of kitchen gloves to keep your hands from getting sticky. Keep squeezing until all the honey is out. At this point the comb will harden into a ball. Break the ball apart and leave it in the strainer. Continue until all your combs are crushed. Leave them to strain until no honey is dripping through the paint strainer. Once all your honey is strained, it's ready for you to jar or enjoy.

Regardless of whether you use the crush-and-strain method or a honey press (see the following section) to extract your honey, the comb is destroyed and the bees will have to build new comb before they can store more honey. That's both a disadvantage and an advantage of Top Bar honey harvesting. The disadvantage is the time and resources that the bees have to expend to draw more comb. The advantage is that you will have fresh, new comb that ensures the purity of your honey products (pesticide buildup can occur on old comb used year after year, as can happen when spinning frames and extracting honey).

Using this simple method, your honey will contain all the goodness the bees put into the honey, including small amounts of pollen and small bits of wax. These tiny particles are suspended in the strained honey but eventually will rise to the top of the container you use. It will look like the honey has a white, foamy coating. It is perfectly healthy for you to enjoy — or it can be removed with a spoon before eating.

**TIP**

Eventually all honey will crystalize. That does not change the purity of the product in any way. If you prefer liquid honey, you can heat the crystalized honey in a warm water bath (120 degrees [49 degrees Celsius]) until it liquefies. Never heat honey excessively.

## Harvesting honey using a honey press

The crush-and-strain method (see the preceding section) of gathering honey from your Top Bar hive is a sticky, messy job. If you have multiple Top Bar–style hives like Warré or Kenyan hives and need to process lots of full honeycombs, the honey press procedure is quite simple:

1.  **One by one, slice the comb off each top bar using a serrated bread knife or an electric uncapping knife.**

    Allow the entire comb (wax and capped honey) to drop into the honey press's container. Depending upon the size and style of your press, you are likely to get several entire combs into the hopper.

2.  **When the hopper is filled, put the lid and screw mechanism in place.**

    Following the manufacturer's instructions (each brand is slightly different), apply pressure (using the screw mechanism; see Figure 17-9) to squeeze and crush the honey from the wax combs. The extracted honey will ooze through the metal or cloth filter, and you can capture it in a clean honey bucket.

3. **Remove the crushed wax, clean the filter, and repeat the process until all the combs have been processed.**

This method for harvesting honey also harvests a lot of beeswax, so be sure to save and process the crushed wax, as explained later in this chapter. See Chapter 18 to learn how to make gifts from your beautiful beeswax.

4. **Return the now-empty top bars to their hive.**

The bees will need to rebuild new comb on these bars before they use them again next season.

FIGURE 17-9: This is one style of honey press that is used for extracting honey from Top Bar hives.

*Courtesy of Swienty Beekeeping*

Honey presses are not that easy to find, and they can be expensive. You are more likely to find honey presses from suppliers in Europe, where they are more commonly used. But if you are clever at building things, the Internet will turn up an array of plans for making your own.

TIP

Swienty, one of the European vendors offering discount coupons at the end of this book, sells a very nice honey press.

## Harvesting cut-comb honey

Cut-comb honey is usually selected from your best-looking comb and never from old or brood comb. With cut-comb you simply cut sections of comb and use them as is. Some folks spread the wax and honey right on the morning toast and others chew small sections like candy. You can store your cut-comb section in plastic containers. I recommend that you freeze the cut comb in containers for 48 hours to kill any wax moth eggs.

# Harvesting Wax

When you extract honey, the cappings that you slice off represent your major wax harvest for the year. You'll probably get 1 or 2 pounds of wax for every 100 pounds of honey that you harvest. If you have a Top Bar hive and are using a honey press, you will have an even greater bounty of beeswax. This wax can be cleaned and melted down for all kinds of uses (see the section on making stuff from beeswax in Chapter 18). Pound for pound, wax is worth more than honey, so it's definitely worth a bit of effort to reclaim this prize. Here are some guidelines:

**1.** **Allow gravity to drain as much honey from the wax as possible.**

Let the wax drain for a few days. Using a double uncapping tank greatly simplifies this process. See Chapter 16 for an image of a double uncapping tank.

**2.** **Place the drained wax in a 5-gallon plastic pail and top it off with warm (not hot) water.**

Using a paddle — or your hands — slosh the wax around in the water to wash off any remaining honey. Drain the wax through a colander or a honey strainer and repeat this washing process until the water runs clear.

**3.** **Place the washed wax in a double boiler for melting.**

WARNING

Always use a double boiler for melting beeswax (never melt beeswax directly on an open flame because it's highly flammable). And *never, ever* leave the melting wax — even for a moment. If you need to go to the bathroom, turn off the stove!

**4.** **Strain the melted beeswax through a couple of layers of cheesecloth to remove any debris.**

Remelt and re-strain as necessary to remove all impurities from the wax.

**5.** **The rendered wax can be poured into a block mold for later use.**

I use an old cardboard milk carton. Once the melted wax has solidified in the carton, it can easily be removed by tearing away the carton. You're left with a hefty block of pure, light-golden beeswax.

# 6

# The Part of Tens

Find out about ancillary bee-inspired hobbies that are a ton of fun, such as planting a bee-friendly garden; brewing mead wine from honey; and making cosmetics, soaps, and candles from the things you harvest from your bees.

Take a look at the answers to the questions I am most frequently asked about honey bees and beekeeping.

Try your hand at cooking up all ten of my favorite honey-inspired recipes.

IN THIS CHAPTER

» **Combining and dividing hives**

» **Building an elevated hive stand**

» **Planting a garden for your bees**

» **Brewing mead: The nectar of the gods**

» **Creating useful gifts from propolis and beeswax**

Chapter **18**

# More than Ten Fun Things to Do with Bees

One of the glorious things about keeping bees is that your interests can expand way beyond the business with your smoker and hive tool. Beekeeping opens up entire new worlds of related hobbies and activities — horticulture, carpentry, biology, and crafts, just to name a few. That's been a good thing for me, because living in Connecticut, as I do, the winters seemed unbearably long when I couldn't be outside playing with my bees. I really missed them! But now, having gotten drawn into some of these related hobbies, I can hardly find time to sit and think. Here are a few of the bee-related activities whose sirens have beckoned to me over the years.

## Making Two Langstroth Hives from One

If you're like most beekeepers I know, it's only a matter of time before you start to ask yourself, "Gee, wouldn't it be twice as much fun to have twice as many hives?" Well, actually it is more fun. And the neat thing is that you can create a second colony from your existing colony. You don't even have to order another package of bees! Free bees!

Ah, but here's the dilemma. You'll need a new queen for your new colony. Strictly speaking, you don't have to order a new queen. You can let the bees make their own; however, ordering a new queen to start a new colony is simply faster and more foolproof. I discuss the nuances of ordering a new queen later in this chapter. Or if you want to get really adventurous, you can raise your own queen (see Chapter 14).

To make two hives from one, you first need a strong, healthy hive. That's just what you hope your hive will be like at the start of its second season — boiling with lots and lots of busy bees. The procedure is known as *dividing* or *splitting* a hive, or *making a divide.*

**TIP**

Dividing not only enables you to start a new colony, it's also considered good bee management — dividing thins out a strong colony and prevents that colony from swarming.

**ALL NATURAL**

Making two hives from one also helps control Varroa mites by hindering mite reproduction. The divide causes a pause in brood production within the new hive. Because Varroa mites reproduce by laying eggs on bee brood, with no new brood available, the mites die out. Read more about this chemical-free way of controlling mite populations in Chapter 13.

The best time to make a divide is in the early spring, about a month before the first major nectar flow. Follow these steps:

1. **Check your existing colony (colonies) to determine whether you have one that's strong enough to divide.**

   Look for lots of bees and lots of capped brood (six or more frames of capped brood and/or larvae are ideal). The colony should look crowded.

2. **Order a new hive setup from your bee supplier.**

   You'll want hive bodies, frames, foundation — the works. You need the elements to build a new home for your new family.

3. **Order a new queen from your bee supplier.**

   Alternatively, you can allow the new colony to raise its own queen. See Chapter 14.

**TIP**

Your new queen doesn't have to be marked, but having a marked queen is a plus, particularly when you're looking for her because the mark makes her easier to identify. I advise you, a new beekeeper, to let your bee vendor mark your queen. A novice can end up killing a queen by mishandling her.

4. **Put your new hive equipment where you plan to locate your new family of bees.**

   You'll need only to put out one deep hive body at this point — just like when you started your first colony (see Chapter 5). Remove four of the ten foundation frames and set them aside. You'll need them later.

5. **When your new marked queen arrives, it's time to divide!**

   Smoke and open your existing colony as usual.

6. **Find the frame with the queen and set it aside in a safe place.**

   An extra empty hive body and cover will do just fine. Better yet, use a small "nuc" hive (available from your supplier). These mini-hives contain only five frames.

7. **Now remove three frames of capped brood (frames with cells of developing pupae) plus all the bees that are on each of them.**

   Place these three brood frames and bees in the center of the new hive. I know, I know — that still leaves one slot open because you removed *four* frames of foundation. The extra slot, however, provides the space that you need to hang the new queen cage (see Step 8).

8. **Using two frame nails, fashion a hanging bracket for the new queen cage (candy side up) and hang the cage between brood frames in the middle of the new hive. Alternatively, if the weather is nice and warm, you can use the bottom-board installation technique (see Chapter 10).**

   Make sure you have removed the cork stopper or metal disc, revealing the candy plug. This is the same queen introduction technique that you used when you installed your first package of bees (see Chapter 6).

9. **Put a hive-top feeder on your new colony and fill it with sugar syrup.**

10. **Turn your attention back to the original hive.**

    Carefully put the frame containing the queen back into the colony. Add three of the new foundation frames (to replace the three brood frames you removed earlier). Place these frames closest to the outer walls of the hive.

11. **Add a hive-top feeder to your original hive and fill it with sugar syrup.**

Congratulations, you're the proud parent of a new colony!

# Making One Langstroth Hive from Two

Keep in mind that it's better to go into the winter with strong colonies — they have a far better chance of making it through the stressful cold months than do weak ones.

If you have a weak hive, you can combine it with a stronger colony. If you have two weak hives, you can combine them to create a robust colony. But you can't just dump the bees from one hive into another. If you do, all hell will break loose. Two colonies must be combined slowly and systematically so the hive odors merge gradually — little by little. This is best done late in the summer or early in the autumn (it isn't a good idea to merge two colonies in the middle of the active swarming season).

My favorite method for merging two colonies is the so-called *newspaper method*. Follow these steps:

1. **Identify the stronger of the two colonies.**

   Which colony has the largest population of bees? Its hive should become the home of the combined colonies. The stronger colony stays put right where it's now located.

2. **Smoke and open the weaker colony (see Chapter 7 for instructions).**

   Manipulate the frames so you wind up with a single deep-hive body containing ten frames of bees, brood, and honey. In other words, consolidate the bees and the ten best frames into one single deep. The "best" frames are those with the most capped brood, eggs, and/or honey.

3. **Smoke and open the stronger hive.**

   Remove the outer and inner covers and put a single sheet of newspaper on the top bars. Make a small slit or poke a few holes in the newspaper with a small nail. This helps hive odors pass back and forth between the strong colony and the weak one that you're about to place on top.

4. **Take the hive body from the weak colony (it now contains ten consolidated frames of bees and brood) and place it directly on top of the stronger colony's hive.**

   Only the perforated sheet of newspaper separates the two colonies (see Figure 18-1).

5. **Add a hive-top feeder and fill it with sugar syrup.**

   The outer cover goes on top of the feeder. No inner cover is used when using a hive-top feeder.

6. **Check the hive in a week.**

   The newspaper will have been chewed away, and the two colonies will have happily joined into one whacking strong colony. The weaker queen is now history, and only the stronger queen remains.

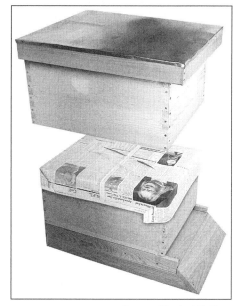

**FIGURE 18-1:**
A weak colony can be combined with a strong colony by using the newspaper method.

*Courtesy of Howland Blackiston*

7. **Now you have the task of consolidating the three deeps back into two.**

   Go through all the frames, selecting the 20 best frames of honey, pollen, and brood. Arrange these in the lower two deeps. Frames with mostly brood go into the bottom deep, and frames with mostly honey go into the upper deep. Shake the bees off the ten surplus frames and into the lower two deeps (save these frames and the third hive body as spares).

# Dividing a Top Bar Hive into Two Colonies

After determining the strength of the colony you want to divide (sometimes referred to as a "split"), either purchase or build another hive body and top bars.

**REMEMBER**

With the exception of the length, the hive dimensions — in terms of the width, depth, and side angle — should be the same for all your Top Bar equipment. This way, bars and drawn comb can be exchanged freely between hives.

Then, purchase a new queen that you will use to re-queen the divide.

Next, determine the location of your new colony. You can place it relatively close to the "parent colony" or locate it some distance away.

If you place the new colony close to the parent, vary the compass direction of the entrance by turning it slightly or making it perpendicular to the parent. This will help the returning foraging bees find their colony and avoid their tendency to drift between colonies that are lined up in a row.

Now select the combs you want to transfer to your new hive. Look over the brood nest in your parent colony and select three combs — one with mostly open brood, one with mostly capped brood, and one that is mostly capped honey. The idea is to take enough eggs, brood, and honey to get your new colony started without setting back the parent colony too far.

One by one, bring the selected combs over to the new hive, leaving three or four blank bars between the entrance and where you place the first comb. Do not remove the bees on these combs — just bring them over full of bees. Keep a sharp lookout for the old queen. Knowing the location of the parent colony's queen is critical, and it means that you must fully inspect each bar before moving it over. If you find the queen on a bar you want to move, put that bar back into the parent colony and select another.

Now, fill in the parent colony's blank spaces with bars you removed from the new colony.

Install the new queen in the new hive (following the procedure in Chapter 6). Fill the feeder and replace the cover.

A quick inspection a few days after the split will help you determine the queen situation. Assuming that's okay, you've made a successful divide and now are the proud owner of a new Top Bar hive and a new colony of bees.

If you end up with the old queen and the caged queen in the same colony, the colony will kill the new queen as soon as she's released from the cage. You will know this happened because inspections will reveal that the colony without a queen is building emergency queen cells to try to raise a new queen.

# Combining Two Top Bar Hive Colonies

In a Top Bar hive, combining two colonies can be accomplished using the newspaper method. First make certain that one of the colonies is queenless by killing the queen in the weaker colony. Then place an empty top bar directly behind the last active comb in the colony where you are adding the weaker colony.

**TIP**

To facilitate installing a blank bar behind your active colony, you can use a follower board. A *follower board* is anything that can be placed inside a Top Bar hive that will stop bees from entering the unused space behind the last occupied bar (see Figure 18-2). If you have a follower board, place it behind the last bar in the colony so you can stop the bees from interfering during the process of combining.

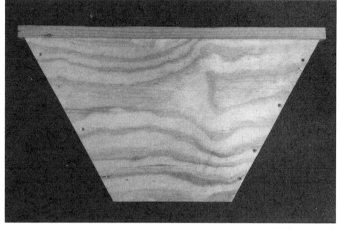

**FIGURE 18-2:**
A follower board stops bees from entering unused space.

*Courtesy of William Hesbach*

Drape a few sheets of dampened newspaper over the blank bar and form it to the sidewalls so no bees can get around it. If you used a follower board, remove it and push the bar with newspaper so it joins the last bar of the colony.

You can now proceed to install the weaker colony into the stronger one by removing each comb from the weaker colony and placing it behind the newspaper-draped bar. The bees from both colonies will eat through the newspaper and, during the time it takes to join, will have acquired a similar scent, thus minimizing fighting. After three or four days you can remove the bar with the newspaper, and you are done.

# Building an Elevated Hive Stand

An elevated hive stand is exactly what it sounds like: an item you use to hold a beehive off the ground. I put all of my hives on this kind of stand. The advantages of using an elevated hive stand are listed in Chapter 5.

**TIP**

If you're reasonably good at woodworking, you can build your own hives and accessories from scratch. My book *Building Beehives For Dummies* (Wiley) includes detailed plans for building six different kinds of hives, plus seven nifty beekeeping accessories.

Figure 18-3 shows plans to help you build your own elevated hive stand. The dimensions of this stand are ideal for a Langstroth hive (eight or ten frame), a Warre hivé, or to hold a couple of five-frame nuc hives. The generous top surface not only accommodates the hive, but there's some extra surface area to place your smoker, tools, and the frames you remove for inspection.

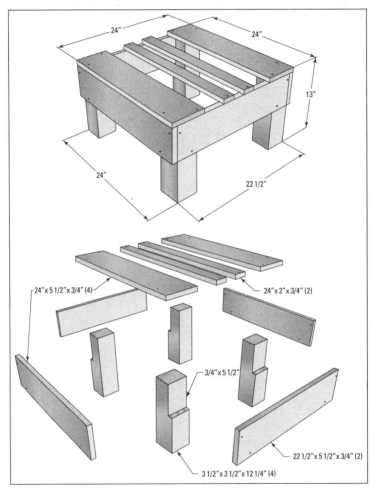

**FIGURE 18-3:** These blueprints serve as a guide for building your own elevated hive stand.

© *John Wiley & Sons, Inc.*

# Building materials list

Table 18-1 lists what you need to build this elevated hive stand.

**TABLE 18-1**     ## Building Materials for Elevated Hive Stand

| Lumber | Hardware | Fasteners |
|---|---|---|
| 2–8′ lengths of 1″ x 6″ cedar or pine lumber | Optional: weatherproof wood glue | 30–6″ x 2″ deck screws, galvanized flat-head Phillips with coarse thread and sharp point |
| 1–8′ length of 4″ x 4″ cedar posts | Optional: a pint of exterior latex or oil paint, exterior polyurethane, or marine varnish to protect the wood from the weather | |

# Cut list

This section breaks down the elevated hive stand into its individual components, and Table 18-2 provides instructions on how to cut and assemble those components.

**TABLE 18-2**     ## Hive Stand Cut-and-Assemble Instructions

| Quantity | Material | Dimensions | Notes |
|---|---|---|---|
| 4 | 4″ x 4″ cedar posts | 12¼″ x 3½″ x 3½″ | These are the leg posts of the stand. Rabbet 5½″ by ¾″ deep along one end of the post (this rabbet accommodates the narrow sides of the stand). |
| 4 | 1″ x 6″ of cedar or pine | 24″ x 5½″ x ¾″ | These are the long sides of the stand and wide struts for the top. |
| 2 | 1″ x 6″ of cedar or pine | 24″ x 2″ x ¾″ | These are the narrow struts for the top. |
| 2 | 1″ x 6″ of cedar or pine | 22½″ x 5½″ x ¾″ | These are the short sides of the stand. |

**TIP**

You can adjust the stand's height to suit your needs by adjusting the length of the 4-x-4-inch cedar posts. Longer legs result in less bending over during inspections. But keep in mind that the higher the stand, the higher your honey supers will be, potentially making it more difficult to lift the heavy, honey-laden supers off the hive. I find the 13-inch height of this design just right for me.

# Planting Flowers for Your Bees

*This section was prepared by my friend Ellen Zampino, an avid gardener and a honey of a beekeeper.*

Flowers and bees are a perfect match. Bees gather nectar and pollen, enabling plants to reproduce. In turn, pollen feeds baby bees, and nectar is turned into honey to be enjoyed by the bees and you. Everyone's happy.

Each source of nectar has its own flavor. A combination of nectars produces great-tasting honey. Not all varieties of the flowers described in the sections that follow produce the same quality or quantity of pollen and nectar, but the ones that I list here work well, and bees simply love them.

While many kinds of trees and shrubs are bees' prime source of pollen and nectar, a wide range of flowers contribute to bee development and a bumper crop of honey. You can help in this process by adding some of these plants to your garden or by *not* removing those that are already there. Did you know that many weeds actually are great bee plants, including the pesky dandelion, clover, goldenrod, and purple vetch?

You can grow annuals such as zinnias, cleome, cosmos, dahlias, and snapdragons by planting seeds directly into the ground or starting them indoors a few weeks before spring. Perennials like ajuga, crane's bill (geranium), campanula, Russian sage, speedwell, and foxglove, as well as many others purchased from local nurseries, provide good sources of pollen and nectar for your bees.

Winter aconite, spring and fall crocus, alliums, and gladiolus are bulbs you can add to your garden that will provide additional beauty as well as a source of pollen and nectar. Lavender, borage, comfrey, thyme, basil, oregano, and garlic chives are just a few of the herbs that provide a great source of nectar and pollen for the bees. They're also great additions to your summer menu. These plants and bulbs can be found in local nurseries or through mail-order catalogs. Blooming time and hardiness vary, depending on your climate. Always make sure you know what you are planting so you don't get a surprise. *Note:* Although comfrey is a long-blooming bee plant, I consider it invasive because it spreads and is almost impossible to contain. Bee warned.

All the plant families listed below are both a good source of pollen and nectar.

## Asters (aster/callistephus)

The *aster family* has more than 100 different species. The aster is one of the most common wildflowers, ranging in color from white and pink to light and dark purple.

Asters differ in height from 6 inches to 4 feet and can be fairly bushy. Asters are mostly perennials, and blooming times vary from early spring to late fall; however, like all perennials, their blooming period lasts only a few weeks. Several varieties can be purchased as seeds, but you can also find some aster plants offered for sale at nurseries.

*Callistephus* are annual China asters, which run the same range of colors but produce varied styles of flowers. These pincushions-to-peony-style flowers start blooming late in summer and continue their displays until frost.

## Bachelor's buttons (Centaurea)

Annual and perennial selections of bachelor's buttons are available. The annuals *(Centaurea cyanus, C. imperialis)*, found in shades of white, pink, yellow, purple, and blue, are also referred to as cornflowers.

The perennial version is a shade of blue or white that blooms in June and sometimes again in late fall. Some gardeners refer to these as mountain blue buttons.

## Bee balm (Monarda)

Bee balm *(Monarda didyma)* is a perennial herb that provides a long-lasting display of pink, red, red purple, and crimson flowers in midsummer. They start flowering when they reach about 18 inches and continue to grow to 3 or 4 feet in height. Deadheading them encourages more growth, which can prolong their flowering period. Bee balm is susceptible to powdery mildew, but the Panorama type does a good job of fending off this problem. Bee balm is a good source of nectar for bees as well as for butterflies and hummingbirds. This family also includes horsemint *(Monarda punctata)* and lemon mint *(Monarda citriodora)*. The fragrant leaves of most of these plants are used in herbal teas.

## Hyssop (Agastache)

Anise hyssop *(Agastache foeniculum)* has a licorice fragrance when you bruise its leaves. The most common form produces tall spikes of purple flowers from midsummer to late fall. There's also a white variety of this plant. Hyssop flowers from seed the first year you plant it, and although not a true perennial, it easily reseeds itself. Another common hyssop is found in the wild — *Agastache nepetoides*. It has a light yellowish flower and is found in wooded areas. While hyssop is a good source of pollen, it's a better source of nectar.

## Malva (Malvaceae)

Malva is a cousin to the hollyhock. I planted zebrina or zebra mallow *(Malva sylvestris)* several years ago and have been a fan ever since. It produces purple flowers that the bees love. It does well in poor weather and produces a wealth of flowers that the bees visit all day long. While not a perennial, it reseeds itself readily and is hardy to zone 5. There are 26 to 30 species in this family, all easily grown from seed.

## Mint (Mentha)

Chocolate, spearmint, apple mint, peppermint, and orange mint are only a few of the types of mints available. Mint flowers are high in nectar. They come in a variety of colors, sizes, fragrances, and appearances, but when they produce a flower, bees are there. Most mints bloom late in the year. Several can be easily grown from seed, while other varieties are obtained from root cuttings. Many gardeners are more than happy to share cuttings because mint is a rapid spreader and extremely hardy. If you don't want it all over your yard, you can contain its growth by planting it in a pot. During the summer, freeze a few leaves along with some honey water and add to your iced tea. Yummy!

## Nasturtium (Tropaeolum minus)

These are annuals grown either from seeds or purchased as plants. These lovely plants come in a wide variety of colors. Many are two-toned. In addition to having beautiful flowers, the Alaska nasturtium has two-tone white and green leaves. These plants range from bushy to trailing and are used at the edges of raised beds, in flower boxes, and hanging baskets. They will bloom until a frost kills them. *Note:* They don't flower well in shade, but can grow to over 4 feet in rich soil.

## Poppy (Papaver/Eschscholzia)

Danish flag *(Papaver somniferum),* corn poppy *(Papaver rhoeas),* and Iceland poppies *(Papaver nudicaule)* are easily grown from seed. Although they can be transplanted, they prefer to be grown from seed. Their colors range from deep scarlet or crimson to various pastel shades. All bloom freely from early summer to fall, need full sun, and grow 2 to 4 feet tall. Some people claim that poppies are valuable mostly for the pollen, but I'm sure my bees also are gathering a fair amount of nectar from poppies.

California poppies *(Eschscholzia)* are golden orange and easily grown. They are a good pollen source for honey bees. California poppies self-seed in warmer climates.

## Salvia (Salvia/farinacea-strata/splendens/officinalis)

The Salvia family, with more than 500 varieties, includes the herb *(Salvia officinalis)*, several native varieties, and many bedding plants. The sages are good nectar providers. When in bloom, they're covered with bees all day long. The variety of colors and sizes of the *farinacea* and *splendens* covers the entire gambit from white, apricot, all shades of red, and purple to blues with bi-colors and tri-colors. Several are perennials; others are annuals.

## Sunflowers (Helianthus/Tithonia)

Sunflowers are made up of two families, and they provide the bees with both pollen and nectar. Each family is readily grown from seed, but you may find some nurseries that carry them as potted plants. When you start sunflowers early in the season, make sure you use peat pots. They are rapid growers that transplant better when you leave their roots undisturbed by planting the entire pot. *Helianthus annuus* include the well-known giant sunflower as well as many varieties of dwarf and multi-branched types. Sunflowers no longer are only tall and yellow. They come in a wide assortment of sizes (from 2 to 12 feet) as well as a range of colors (from white to rust). There are several varieties with a mixture of colors. Watch out for the hybrid that is pollenless because it is of little use to the bees. Several varieties will easily self-seed, even in the cold Northeast.

# Brewing Mead: The Nectar of the Gods

I get restless every winter when I can't tend to my bees. So a number of years ago I looked around for a related hobby that would keep me occupied until spring. I thought, "Why not brew mead?" Mead is a wine made from honey instead of grapes. It was the liquor of the Greek gods and is thought by scholars to be the oldest form of alcoholic beverage. In early England and until about 1600, mead was regarded as the national drink. In fact, the wine that Robin Hood took from Prince John had honey as its base.

When mead is made right, the resulting product is simply delicious! And like a fine red wine, it gets better and better with age. Many bee supply vendors supply basic wine- and mead-making equipment to hobbyists (see Figure 18-4). All you need is a little space to set up shop and some honey to ferment. The key to success is keeping everything sanitary — sterile laboratory conditions!

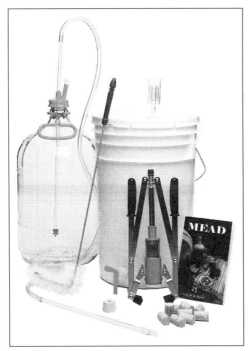

FIGURE 18-4:
Here's a typical kit
for brewing mead
(honey wine).

*Courtesy of Howland Blackiston*

The following recipe produces an extraordinary mead. Technically, this is a *Metheglin*, the term given to mead that is spiced. The recipe yields about 40 750-ml bottles of finished product. Adjust the amounts to suit your needs.

**TIP**

Ideally, keep the room's temperature between 65 and 68 degrees Fahrenheit (18 and 20 degrees Celsius; a cool basement is a good place to brew mead). If the temperature is higher than 75 degrees (24 degrees Celsius), the yeast may die; if it's lower than 50 degrees (10 degrees Celsius), fermentation ceases. Note that a portable space heater with a thermostat helps control basement temperatures during winter.

1. **The initial honey and water mixture is called the *must*. My recipe calls for the following:**

   - 32 pounds of dark wildflower honey

   - 5 gallons of well water (nonchlorinated water)

   - 5 sticks of cinnamon

   - 1 tablespoon cloves

2. **Add the following to the must:**

   - 4¼ tablespoons of wine yeast nutrient (available at wine-making supply stores)

3. **Pour the mixture into a large (16.5-gallon) initial fermentation tank.**

   Top off with water so the tank contains a total of 13 gallons of must. Stir vigorously to blend and introduce oxygen (splashing permitted).

4. **Add the following ingredients to the tank of must:**

   - 13 potassium metabisulfite tablets (available at wine-making supply stores) to hinder the growth of undesirable bacteria

   - A few drops of antifoam agent (available at wine-making supply stores)

5. **Wait 24 hours and then add the following to the must in the fermentation tank:**

   - 2½ packets of white wine yeast (stir to blend)

6. **Cover loosely and let the must ferment for three to four weeks before performing the first racking (when the bubbling and fizzing has stopped, it's time to rack).**

   *Racking* is the process of siphoning off the liquid and leaving the dead yeast cells behind.

7. **After the initial three to four weeks, rack liquid into glass carboys (large glass vessels). You'll need two or three 5-gallon carboys for this recipe.**

   Fill right up to the neck of the carboy (you want to minimize air space). Add one potassium metabisulfite tablet for each gallon of liquid to maintain 50 parts per million (ppm). Place a fermentation valve on each. The valve keeps air and bacteria from entering the carboy.

8. **Rack a total of two or three more times at one- to two-month intervals.**

   Each racking further clears the mead. I know you're eager to drink your mead, but your patience will pay off in a product that tastes great and has great eye appeal. After the final racking, transfer the mead to sterilized wine bottles and cork tightly. Store bottles on their sides in a cool, dark place. Remember, the longer the mead is aged, the more improved the flavor. Salute!

**TIP**

You can get your wine sparkling clear by using a special filtering device available from beer- and wine-making suppliers.

For more information on making mead, see *Making Mead Honey Wine: History, Recipes, Methods and Equipment* by Roger A. Morse (Wicwas Press).

# Create Cool Stuff with Propolis

*Propolis* (sometimes called "bee glue") is the super-sticky, gooey material gathered by the bees from trees and plants. The bees use this brown goop to fill drafty cracks in the hive, strengthen comb, and sterilize their home.

Propolis has remarkable antimicrobial qualities that guard against bacteria and fungi. Its use by bees makes the hive one of the most hygienic domiciles found in nature. This remarkable property has not gone unnoticed over the centuries. The Chinese have used it in medicine for thousands of years. Even Hippocrates touted the value of propolis for healing wounds. In addition, propolis has been used for centuries as the basis for fine wood varnishes.

When cold, propolis is hard and brittle. But in warm weather, propolis is gummier than words can express. When you inspect your hives at the end of the summer and early autumn (the height of propolis production), you'll discover that the bees have coated just about everything with propolis. The frames, inner cover, and outer cover will be firmly glued together, and they'll require considerable coaxing to pry loose. You'll get propolis all over your hands and clothes, where it will remain for a long, long time. It's a nuisance for most beekeepers. But be sure to take the time to scrape it off, or you'll never get things apart next season. Be sure to save the propolis you scrape off with your hive tool! It's precious stuff. I keep an old coffee can in my toolbox and fill it with the propolis I remove from the hive. And I keep another can for the beeswax (burr comb) I remove.

**TIP**

Keep a spray bottle of rubbing alcohol in your supply box. Alcohol works pretty well at removing sticky propolis from your hands. But, for goodness' sake, keep propolis off your clothes because it's nearly impossible to remove.

Many beekeepers encourage the bees to make lots of propolis. Special *propolis traps* are designed just for this purpose. The traps usually consist of a perforated screen that is laid across the top bars — similar to a queen excluder, but the spaces are too narrow for bees to pass through (see Figure 18-5). Instinctively, bees fill all these little holes with propolis. Eventually, the entire trap becomes thickly coated with the sticky, gummy stuff. Remove the trap from the hive (gloves help keep you clean) and place it in the freezer overnight so the propolis becomes hard and brittle. Like chilled Turkish Taffy, a good whack shatters the cold propolis, crumbling it free from the trap. It then can be used to make a variety of nifty products. I've included some recipes to get you started.

**FIGURE 18-5:**
A propolis trap can be placed where the inner cover usually goes. In no time, bees will coat the entire trap with precious propolis.

## Propolis tincture

Here's a homemade and all-natural alternative to iodine. *Note:* Like iodine, it stains. Use it on minor cuts, rashes, and abrasions. Some folks even use a few drops in a glass of drinking water to relieve sore throats.

1. **Measure the crumbled propolis and add an equal measure of 100-proof vodka or grain alcohol (for example, 1 cup propolis and 1 cup alcohol). Place in an ovenproof bottle with a lid.**

2. **Heat the closed bottle in a 200-degree (Fahrenheit) oven. Shake the bottle every 30 minutes. Continue until the propolis has completely dissolved in the alcohol.**

3. **Strain the mixture through a paper coffee filter or a nylon stocking.**

4. **Bottle the tincture into dropper bottles, which you can get from your pharmacist.**

## Propolis ointment

This ointment can be applied to minor cuts, bruises, and abrasions.

1. **Melt the ingredients in a microwave or in a double boiler.**

   - 1 teaspoon of beeswax

   - 4 teaspoons of liquid paraffin

   - 1 teaspoon of finely chopped propolis granules

   - 1 teaspoon of honey

2. **Remove from heat and stir continuously until it cools and thickens.**

3. **Pour into suitable jars.**

## Propolis varnish

If you happen to have a multimillion-dollar violin made by Stradivarius, you already know that the finest string instruments ever made had a varnish made from propolis. But this superior lacquer need not be reserved for such exclusive uses. Propolis varnish provides a warm, durable finish for any wood project. Here's a recipe from a friend of mine who refinishes museum-quality violins.

1. **Combine all ingredients in the following list in a glass jar at room temperature. Cover the jar with a lid. Allow mixture to stand for a week or more while shaking once or twice a day.**

   - 4 parts blond shellac

   - 1 part manila copal (a soft resin)

   - 1 part propolis

2. **Filter solution through a few layers of cheesecloth or a nylon stocking before using.**

*Note:* The manila copal resin is available from specialty varnish suppliers, such as Hammerl GmbH & Co. KG, Hauptstrasse 18, 8523 Baiersdorf, Germany.

# Making Gifts from Beeswax

Your annual harvest doesn't begin and end with honey. You'll also be collecting plenty of beautiful, sweet-smelling beeswax, which can be cleaned and used for all kinds of artsy projects (see Chapter 15 for instructions on how to clean your wax). You can make candles, furniture polish, and cosmetics for all your friends and neighbors (see Figure 18-6). Better yet, why not sell these goodies at the local farmers' market? Here's some useful information to get you started.

## Beeswax candles

Beeswax candles are desirable; unlike paraffin, they don't drip, don't sputter, and don't smoke, but they do burn a long time. You can make three basic types of candles from beeswax: rolled, dipped, and molded. Buy them in a gift store, and they're fantastically expensive. But not when you make them yourself!

**FIGURE 18-6:**
Here's a sample of the wonderful products you can make from beeswax.

*Courtesy of Howland Blackiston*

## Dipped candles

This is a time-consuming process, but the end result is beautiful.

1. **Melt beeswax in a tall container (the container can be placed in a hot water bath to keep the wax melted).**

   **WARNING**

   *Never* melt beeswax directly over a heat source; *always* use a water bath when melting beeswax. At temperatures higher than 200 degrees (93 degrees Celsius), beeswax can vaporize and ignite.

2. **Tie a lead fishing weight to one end of wicking (to make it hang straight) and begin dipping.**

3. **Let each coat of wax cool before dipping again. The more you dip, the thicker the candle becomes.**

With a little finesse, you can create an attractive taper to your dipped candles. You can even add color and scent (your candle-making supplier sells what you need, including wicks, coloring dyes, and scents). Elegant!

## Molded candles

Candle-making suppliers offer a huge variety of rubber or plastic molds for candle making — from conventional tapers to complex figurines. Just melt your beeswax and pour it into the mold (add color and scent if you want). Don't forget the wick. Let it cool and remove the mold. Easy!

## Beeswax furniture polish

My good friend Peter Duncan makes simply beautiful wood furniture. He says that my beeswax wood polish is the finest he's ever used. Smooth enough to apply evenly, beeswax polish feeds and preserves the wood and provides a hard protective finish. Here's my "secret" recipe:

1. **Gather the following ingredients:**

   - 4 ounces beeswax (by weight)

   - 2 tablespoons of carnauba wax flakes

   - 2½ cups odorless turpentine or mineral spirits

2. **Melt the waxes in a double boiler.**

3. **Remove the waxes from the heat and stir in the turpentine or mineral spirits.**

4. **Pour into containers (something that looks like a tin of shoe polish is ideal) and let the mixture cool.**

5. **Cover tightly with a lid.**

Apply the polish with a clean cloth and rub in small circles. Turn the cloth as it becomes dirty. Allow the polish to dry, then buff with a clean cloth. If more than one coat is desired, wait two days between applications. This stuff is simply fantastic!

# Beauty and the Bees

*Special thanks to my friend and fellow beekeeper Patty Pulliam for putting together these great recipes.*

There is nothing more satisfying than making your own creams, salves, and balms with beeswax from your hives. It's been said that Cleopatra's beauty regime included a cream made with beeswax and olive oil, and some of my favorite recipes use both of these ingredients. Talk about enduring the test of time! Honestly, the best thing about making your own body products is that you know what is in them. No need to worry about parabens or endocrine-disrupting chemicals or ingredients that you can't even pronounce. The beauty products that you can make on your own are far better and fresher than any product out there and are more emollient and hydrating than anything you can buy.

# Use your cappings

In making beauty products, I suggest you only use *cappings wax*. This is the wax that the bees produce to cover the honey in the hive and is saved during extraction. Because it is brand-new, the wax is light in color, aromatic, and clean. And because you already know that no chemicals should be used while the honey supers are on the hive, you're guaranteed that your wax is also chemical-free.

# Equipment

I use a good quality kitchen scale that measures in ounces for the dry weight of beeswax; for the liquid measurements, I use standard baking spoon measurements and measuring cups. To break up the beeswax, use a screwdriver and a hammer on a cutting board to chisel the wax into pieces — *never* use a knife!

Cleaning any container that has been used to melt beeswax can be a lot of work, so I use empty metal coffee cans (well washed and dried, naturally) with chopsticks or the wooden paint stirrers from a hardware store as my mixers. If you don't have a metal coffee can, go to your favorite pizza restaurant and ask for an empty tomato sauce can. Melting wax in a can works very well using a double boiler approach — you simply place the can in a pot of warm water to melt the wax and blend in the oil; bend the lip of the can with a pair of pliers to make a pouring spout. Clean up by wiping the can with paper towels or recycle the can. For a thermometer, I employ a digital-display probe type.

**WARNING**

Working with melted beeswax is like working with hot cooking oil — use common sense! *Never* melt beeswax directly over a heat source; *always* use a water bath when melting beeswax. Wax melts at between 143 and 148 degrees (62 and 64 degrees Celsius). It's fairly stable below 200 degrees (93 degrees Celsius), but at higher temps beeswax can vaporize (flash point) and ignite.

Keep a fire extinguisher handy when melting beeswax. You may also want to cover your countertop with newspaper to absorb any spilled beeswax. This is especially helpful when filling lip balm tubes.

**WARNING**

As wax overheats, it begins to smoke and give off an acrid smell. If this happens, remove it immediately from heat.

# The recipes

The recipes here use extra virgin olive oil, sweet almond oil, and coconut oil — all of which are edible. I figure that I don't want to put anything on my body that I would not put in it. There are so many other wonderful oils, each with its own unique quality, such as shea nut butter, jojoba oil, apricot kernel oil, grapeseed oil, safflower oil, and wheat germ oil, that I encourage you to give them a try. You can substitute any oil here for another; however, I do not suggest that you use mineral or baby oils because they produce a heavy product and are by-products of petroleum production.

When essential oils are called for in the recipes, I encourage you to use one that is pleasing to you. An essential oil is a liquid that's generally distilled (most frequently by steam or water) from the leaves, stems, flowers, bark, roots, or other elements of a plant. Essential oils contain the true essence of the plant from which it was derived. Essential oils are highly concentrated, and a little goes a long way. Essential oils can vary greatly in quality and price; the reasons are generally the rarity of the plant and where it's grown. Essential oil is not the same as perfume or fragrance oils.

Essential oils should be added after the beeswax and oil are melted and off of the heat; they are volatile, and you do not want them to evaporate! One of my favorite uses of essential oils is in the Beeswax and Olive Oil Salve (recipe to follow), which uses different essential oils to make a variety of salves for different purposes. But the ultimate is to use my favorite essential oils in this salve to make layering scents. I make a batch of the salve and then divide it into four (or more, however many you want) batches; then I put a single essential oil in each one. When I apply them as perfume, I can choose to create my own personal scent, depending on my mood.

Because these are fresh and natural cosmetic recipes, they don't contain any preservatives. I encourage you to mark them with the date of manufacture and to use them within a six-month period — but I don't think that will be a problem.

WARNING

Be certain that you use chemically pure borax when making cosmetics (it can be ordered from beekeeping equipment suppliers). Never use laundry-grade borax for cosmetics. Borax is a complex borate mineral salt, which occurs naturally and is mined in the Mojave Desert in Boron, California. In these recipes, it acts as an emulsifier and has some minor preservative properties.

## Ultra-rich skin cream

This cream contains lanolin, which comes from wool of domestic sheep breeds and is used in many dermatological products for compromised skin conditions. It is very emollient and great for dry, scaling, cracked skin.

1. **Gather the following ingredients:**

   - 2½ ounces (weight) beeswax

   - 4 ounces (liquid) lanolin

- ⅔ cup sweet almond oil

- ¾ cup distilled water

- 1 teaspoon borax (sodium borate, chemically pure)

- A few drops essential oil (your choice — I like rose or citrus)

2. **In a double boiler, melt the oil, lanolin, and beeswax to 160 degrees.**

3. **Heat the borax and water in a separate container to 160 degrees. Make sure the borax is dissolved.**

4. **Add the water mixture to the oil mixture while stirring briskly.**

5. **When white cream forms, stir slowly until the mixture cools to 100 degrees.**

6. **Pour into containers, label, and date recipe (use product within six months).**

## Rich body balm

1. **Gather the following ingredients:**

- 5 ounces beeswax (weight)

- 1⅓ cups distilled water

- 2 teaspoons borax

- 2 cups (16 ounces) olive oil

- A few drops of essential oil (your choice — geranium is nice!)

2. **In a double boiler, melt the beeswax and stir in the oils. Heat the water and borax in a separate container to 160 degrees.**

3. **Add the water to the oil mixture *very slowly,* stirring constantly.**

4. **When mixture is emulsified, pour into containers, label, and date the recipe. Use product within six months.**

## Beeswax lip balm

Here's a recipe to make lip balm that'll keep your lips soft and healthy, even in the harshest weather. You can add a bit of color to tint the balm. The tint can be purchased from many of the Internet beauty links listed at the end of this section.

1. **Gather the following ingredients:**

- 1 ounce (weight) beeswax

- 4 ounces (volume) sweet almond oil

- A few drops essential oil (I recommend peppermint or wintergreen)

2. In a double boiler, melt the beeswax and stir in the sweet almond oil.

3. Remove from heat, add a few drops of essential oil, and pour into lip balm containers.

4. Let lip balm cool and solidify before placing caps on containers.

## Beeswax and olive oil salve

TIP

This recipe is a favorite for making specialty salves. By changing the essential oil, you can come up with many different products. Add eucalyptus oil and you have a chest cold remedy. Add comfrey and you have a cut and wound healer. Add propolis and you have a first-aid antibiotic ointment. Add chamomile for a soothing foot-rub salve. Or add citronella or lemongrass for an effective insect repellant.

1. Gather the following ingredients:

   - 1 part beeswax

   - 6 parts olive oil

   - Fresh or dried herbs (optional) or essential oil of your choice

TIP

   If using herbs, clean and dry thoroughly, place in glass jar, cover with olive oil, and allow everything to steep for one week. Strain herbs from olive oil and proceed.

2. In the top of a double boiler set over medium heat, warm olive oil and add beeswax; stir until beeswax is dissolved. Add the optional dried herbs or essential oils, and stir for a minute or two.

3. While still warm, pour into small jars; when cool, cover with lids.

## Beeswax lotion bar

These bars are great to keep handy when you need to refresh your skin. I like to use silicone baking molds, which are available in cookware departments and come in a variety of shapes. Soap-making forms also work nicely.

1. Gather the following ingredients:

   - 2 ounces (weight) beeswax

   - 2 ounces (weight) sweet almond oil

   - 2½ ounces (weight) coconut oil or cocoa butter or combination

   - ¼ teaspoon vitamin E oil

   - A few drops of essential oil (your choice — lavender is lovely!)

2. **Melt all ingredients (except essential oil) in double boiler.**

3. **Remove from heat and add the essential oil.**

4. **Pour into molds, let cool, and place in cellophane wrapper or reusable container.**

## Natural homemade sunscreen

This one is unlike the other finished recipes in that the final product looks more stirred or whipped rather than flat and finished. The zinc oxide is heavier than the other ingredients. To ensure the emulsion is suspended, you need to continue stirring until cool.

Use as you would regular sunscreen. The product is best when used within six-months.

1. **Gather the following ingredients:**

   - ½ cup almond or olive oil (you can infuse with herbs first if desired)

   - ¼ cup coconut oil (natural SPF 4)

   - 2 ounces beeswax

   - 2 tablespoons zinc oxide (available in the pharmacy section of your grocery or a drugstore)

   - Optional: 1 teaspoon vitamin E oil

   - Optional: 2 tablespoons shea butter (natural SPF 4–5)

   - Optional: Essential oil of your choice

2. **Melt all ingredients (except zinc oxide and essential oil if using) in double boiler until melted.**

3. **Remove from heat and add zinc oxide and essential oil; stir until blended.**

4. **Pour into container, stir a few times as it cools to make sure zinc oxide is remaining in suspension.**

# Packaging and labeling

There are many online sites for packaging and labeling products. Think up a clever name for the item you have prepared. Make sure you list all the ingredients on the label (in descending order of quantity) along with the net weight of the product and some sort of contact information. We know that all the products from the hive are amazing and health-giving; just remember that you should not make any guarantees about the benefits of what your creams, salves, and balms will provide. Just include anecdotal testimonials.

I make sure the containers I order are food-grade quality; metal tins are very cute, but some essential oils can react with them and corrode the metal.

Here are some sources for containers, labels, and ingredients:

www.sks-bottle.com/

www.wholesale.glorybee.com

www.thesage.com

www.beehivebotanicals.com

www.wholesalesuppliesplus.com

www.labelsonline.com

» Understanding *unusual* bee behavior

» Addressing concerns specific to Top Bar hives

# Chapter **19**

# More than Ten Frequently Asked Questions about Bee Behavior

am on the receiving end of virtually every question you can possibly imagine about bees and beekeeping. I've received many emails and letters since writing the first three editions of *Beekeeping For Dummies.* Beginning beekeepers face all kinds of puzzling new situations and concerns every day. I know how gratifying it is for them to have someone they can ask when they just can't seem to figure out what to do next. I had a wonderful mentor when I started beekeeping, and it made all the difference when I encountered something baffling.

Not surprisingly, most new beekeepers face the same bewildering situations and ask identical questions. That gives me the illusion of intelligence when I rattle off lucid answers to seemingly impossible conundrums. Actually, it's just that I've had a lot of practice responding to the same questions again and again. The good thing is that I get a little better at it each time.

This chapter includes some of the most frequently asked questions about bee behavior that come my way. They may solve a riddle or two for you as you embark on the wonderful adventure of backyard beekeeping. Look over these questions and answers. Return to this chapter if you encounter a puzzling question down the road.

*Help! A million bees are clustered on the front of my hive. They've been there all day and all night. Are they getting ready to swarm?*

They're not swarming. Chances are it's hot and humid, and the bees are doing just what you'd do — going out on the front porch to cool off. It's called *bearding.* They may spend days and nights outside the hive until the weather becomes more bearable inside. Make sure you've given them a nearby source for water and provided adequate hive ventilation. Bearding can be an indication that the hive's ventilation is not what it should be. (See Chapter 9 for more on ventilation.)

*Is something wrong with my bees? They're standing at the entrance of the hive, and it looks like they're just rocking back and forth. Are they sick?*

Your bees are fine. They're scrubbing the surface of the hive to clean and polish it. They do this inside and outside the hive. Tidy little creatures, aren't they?

*I hived a new package of bees last week. I just looked in the hive. The queen isn't in her cage, and I don't see her or any eggs. Should I order a new queen?*

It's probably too early to conclude that you have a problem. Overlooking the queen is easy (she's always trying to run away from the light when you open a hive). Seeing eggs is a far easier method of determining whether you have a queen. But it may be too soon for you to see eggs. Give it another few days and then look again for eggs. Until they get a better idea of what eggs look like, most new beekeepers have a hard time recognizing them. Have a close look at the images of eggs in the color insert section and in Chapter 2.

A few days after the queen lays the eggs, they hatch into larvae, which are easier to see than eggs. If you see absolutely nothing after ten days (no queen, no eggs, and no larvae), order a new queen from your beekeeping supplier.

*Why is my queen laying more than one egg in each cell? Is she just super productive?*

Sometimes, for a short period of time, a new, young queen may lay more than one egg in the cells. This is normal and of no concern.

However, if you see multiple eggs in many cells over a period of time, you may have a problem. It likely indicates that you have lost your queen, and some of the young worker bees have started laying eggs — a situation referred to as *drone-laying workers.*

If you have drone-laying workers, you'll have to remove them from the hive and get a new queen (see Chapter 10). If you don't correct the situation, you'll eventually lose your hive as all the worker bees die off from old age. At that point only drones are left. Without the workers, there will be no bees gathering food and no workers to feed the helpless drones.

*Hundreds of bees are around my neighbor's swimming pool and birdbath. The bees are creating a problem, and the neighbor is blaming me. What can I do?*

Bees need lots of water in summer to cool the colony, and your neighbor's pool and birdbath are probably the bees' closest sources. You must provide your bees with a closer source of water (see Chapter 3). If they're already imprinted on your neighbor's oasis, you may have to "bait" your new water source with a light mixture of sugar water. After the bees find your sweet new watering hole, you can switch to 100 percent water.

*A tremendous amount of activity is present at the entrance of the hive. It looks like an explosion of bees flying in and out of the hive. The bees seem to be wrestling with each other and tumbling onto the ground. They appear to be fighting with each other. What's going on?*

It sounds like you have a robbing situation. Your bees are trying to defend the hive against invading bees that are stealing honey from your hive. You must call a halt to this activity before the robbing bees steal all the honey and many bees die in the battle (see Chapter 10 for tips on how to prevent and solve this situation).

*My bees had been so sweet and gentle, but now I'm scared to visit the hive. They have become unbearably aggressive. What can I do?*

Bees become more aggressive for a number of different reasons. Consider the following possibilities, and see whether any apply to your situation:

>> A newly established colony almost always starts out gentle. As the colony grows in size and the season progresses, the bees become more protective of their honey stores. Likewise, a growing colony means many more bees for you to deal with. But if the colony is handled with care, this is seldom a problem. Be gentle as you work with your colony.

>> When there is little nectar and pollen available for foraging, the bees can become very possessive and defensive — especially in the autumn.

Incorrect use (or lack of use) of the smoker can result in irritable colonies. See Chapter 7 for information about how to use your smoker.

- » Do you launder your bee clothes and veil? Previous stings on gloves and clothing can leave behind an alarm pheromone that can stimulate defensive behavior when you revisit the hive. Be sure to keep your garments clean. You can also smoke the area of the sting to disguise any alarm pheromone that may linger on clothing or on your skin.

- » When colonies are raided at night by skunks or other pirates, they can become cross and difficult to deal with. See Chapter 13 for ideas on how to remedy these situations.

- » Do you still have your original queen? Are you sure? If you had a marked queen, you'd know for certain whether the queen now heading your colony is your original queen (see if she's marked!). A colony that supersedes the queen sometimes can result in more aggressive bees. That's because you have no guarantee of the new genetics. The new queen mated with drones from goodness knows where. Her offspring may not be as nice as the carefully engineered genetics provided by your bee supplier. When this happens, order a marked and mated queen from your supplier to replace the queen that is now in your hive.

When you purchase a marked queen from a supplier, the marking stays on for the full life of the queen. It's like spilling nail polish on the living room carpet. It never wears off!

*I see white spots on the undersides of my bees. I'm worried these might be mites or some kind of disease. What are these white flecks?*

This isn't a problem. The white flakes that you see are bits of wax produced by glands on the underside of the bee's abdomen. They use this wax to build comb. All is well.

*The bees have carried dead larvae out of the hive and dumped them in and around the entrance of the hive. What's going on?*

Bees remove any dead bees and larvae from the hive. They keep a clean house. Sometimes the dead larvae may be *chilled brood*, or brood that died when the temperature took a sudden and unexpected drop. Larvae that look hard and chalky may be a sign of chalkbrood (see Chapter 12 for more information on chalkbrood). Either case is fairly commonplace. You don't need to be concerned unless the number of dead bees and larvae is high (more than ten).

*It's midwinter, and I see quite a few dead bees on the ground at the hive's entrance. Is this normal?*

Yes. Seeing a few dozen dead bees in and around the hive's entrance during the winter months is normal. The colony cleans house on mild days and attempts to remove any bees that have died during the winter. In addition, some bees may take "cleansing flights" on mild sunny days but may become disoriented or caught in a cold snap. When that's the case, they don't make it all the way back to the hive — dropping dead in the snow, sometimes near the entrance of the hive. Seeing more than a few dozen dead bees may be an indication of a health problem, so it may be time for a closer inspection on the first mild, sunny day.

*I see some bees with shriveled wings and very short, stubby abdomens. Are these just baby bees?*

These are not baby bees. This is usually a warning sign that the colony has a virus epidemic due to a significant infestation of Varroa mites. This results in malformations, like shortened abdomens, misshapen wings, and deformed legs. It's time to take action. See Chapter 13 for more about Varroa mites and what you can do about them.

*Since my Top Bar hive is a more "natural" way to keep bees, can I skip the monitoring and treatment for mites and other diseases?*

Having a Top Bar hive doesn't mean you can ignore your bees. You should consider the same monitoring and potential treatment for mites and other diseases. It's all part of being a responsible beekeeper.

*Is my Top Bar hive legal?*

Yes, it is. Top Bar hives are considered a "removable frame hive" that can be fully inspected by the local authorities. Therefore they are legal in all 50 states and Canada.

*Because Top Bar hives were developed for use in Africa, will my Top Bar colony do okay in a colder climate?*

Colonies that make their home in Top Bar hives have the same chance of surviving cold weather as any other colony in any other type of hive. The key to cold-weather survival is strong, healthy bees and a large population with ample honey going into the winter.

# Chapter **20**

# My Ten Favorite Honey Recipes

There's a good chance that you'll be able to harvest 40 or more pounds of honey from each of your hives. That's a lot of honey. Unless you eat a whole lot of toast, you may want to consider other ways to use your copious crop. Honey is not only wholesome, delicious, sweet, and fat-free, but it's also incredibly versatile. You'll find uses for honey in a myriad of recipes that call for a touch of sweetness.

In this chapter, I include ten of my favorite recipes from the National Honey Board. For additional recipes, be sure to visit its website (www.honey.com) or write to the National Honey Board, 11409 Business Park Circle, Ste. 210, Firestone, CO 80504-9200; phone 303-776-2337.

The recipes here cover appetizers, baked goods, main dishes, and desserts. Before I jump into the recipes themselves, here are some tips for cooking with honey:

» Because of its high fructose content, honey has a higher sweetening power than sugar. This means you can use less honey than sugar to achieve the desired sweetness.

» To substitute honey for sugar in recipes, start by substituting up to half of the sugar called for. With a little experimentation, honey can replace all the sugar in some recipes.

» If you are measuring honey by weight, 1 cup of honey will weigh 12 ounces.

» For easy cleanup when measuring honey, coat the measuring cup with nonstick cooking spray or vegetable oil before adding the honey. The honey will slide right out.

» In baking, honey helps baked goods stay fresh and moist longer. It also gives any baked creation a warm, golden color. When substituting honey for sugar in baked goods, follow these guidelines:

- Reduce the amount of liquid in the recipe by ¼ cup for each cup of honey used.

- Add ½ teaspoon of baking soda for each cup of honey used.

- Reduce the oven temperature by 25 degrees to prevent overbrowning.

# Honey Curry Vegetable Dip

**PREP TIME: 5 MIN | YIELD: APPROXIMATELY 1 CUP**

**INGREDIENTS**

1 cup low-fat mayonnaise

¼ cup honey

1 tablespoon curry powder

1 tablespoon white wine vinegar

Assorted fresh raw vegetables (celery, carrots, cauliflower, broccoli)

**DIRECTIONS**

1 Combine mayonnaise, honey, curry, and vinegar; mix well.

2 Refrigerate about 1 hour to allow flavors to blend. Serve as a dip with vegetables.

**PER SERVING (PER TABLESPOON):** *Calories: 37 (From Fat 44%); Fat: 2g; Cholesterol: 8mg; Sodium: 18mg; Carbohydrate: 5.3g (Dietary Fiber 0.1g); Protein: 0.3g.*

# Golden Cornbread

PREP TIME: 15 MIN | COOK TIME: 25 MIN | YIELD: 8 SERVINGS

**INGREDIENTS**

3 cups yellow cornmeal

1 cup whole-wheat flour

2 tablespoons baking powder

1 teaspoon salt

2 cups buttermilk or low-fat yogurt

½ cup butter or margarine, melted

½ cup honey

3 eggs, beaten

**DIRECTIONS**

1 Combine cornmeal, flour, baking powder, and salt in large bowl.

2 Combine buttermilk, butter, honey, and eggs in separate large bowl.

3 Stir buttermilk mixture into flour mixture just until moistened.

4 Pour into greased 12-x-8-x-2-inch baking pan. Bake at 350 degrees for 25 minutes or until golden brown. Let cool and cut into 8 squares.

**PER SERVING (PER SQUARE):** *Calories: 552 (From Fat 27%); Fat: 17g; Cholesterol: 132mg; Sodium: 584mg; Carbohydrate: 8g (Dietary Fiber 9g); Protein: 19g.*

# Honey Picante Chicken Wings

**PREP TIME: 10 MIN | COOK TIME: 50 MIN | YIELD: 24 TO 30 SERVINGS**

**INGREDIENTS**

½ cup honey

½ cup prepared picante sauce

2 cloves garlic, minced

¼ teaspoon Worcestershire sauce

2 pounds chicken wings, tips removed

Salt and pepper, to taste

**DIRECTIONS**

1 Microwave honey in a 1-quart microwave-safe bowl, on high (100%) 4 to 5 minutes or until honey boils and thickens.

2 Stir in remaining ingredients except chicken wings; microwave on high 4 to 5 minutes.

3 Cut wings in half at joint; arrange wings in shallow baking dish.

4 Bake at 350 degrees for 15 minutes. Raise temperature to 375 degrees.

5 Turn each piece, brush generously with sauce in pan and bake 45 minutes longer or until glazed and completely cooked, turning chicken and brushing with sauce every 15 minutes.

**PER SERVING (ONE WING):** *Calories: 118 (From Fat 50%); Fat: 6g; Cholesterol: 26mg; Sodium: 152mg; Carbohydrate: 6g (Dietary Fiber 0g); Protein: 9g.*

# Apricot Honey Bread

**PREP TIME: 20 MIN | COOK TIME: 60 MIN | YIELD: 12 SERVINGS**

**INGREDIENTS**

3 cups whole-wheat flour

3 teaspoons baking powder

1 teaspoon ground cinnamon

½ teaspoon salt

¼ teaspoon ground nutmeg

1¼ cups 2% low-fat milk

1 cup honey

1 egg, lightly beaten

2 tablespoons vegetable oil

1 cup chopped, dried apricots

½ cup chopped almonds or walnuts

½ cup raisins

**DIRECTIONS**

1 Combine the flour, baking powder, cinnamon, salt, and nutmeg in a large bowl and set aside.

2 Combine the milk, honey, egg, and oil in a separate large bowl.

3 Pour the milk mixture over the dry ingredients and stir until just moistened.

4 Gently fold in the apricots, nuts, and raisins.

5 Pour into a greased 9-x-5-x-3-inch loaf pan. Bake at 350 degrees for 55 to 60 minutes or until a wooden pick inserted near the center comes out clean.

**PER SERVING:** *Calories: 302 (From Fat 15%); Fat: 6g; Cholesterol: 20mg; Sodium: 154mg; Carbohydrate: 61g (Dietary Fiber 5g); Protein: 7g.*

# Asian Honey-Tea Grilled Prawns

**PREP TIME: 15 MIN | COOK TIME: 6 MIN | YIELD: 4 SERVINGS**

**INGREDIENTS**

1 cup brewed double-strength orange spice tea, cooled

¼ cup honey

¼ cup rice vinegar

¼ cup soy sauce

1 tablespoon fresh ginger, peeled and finely chopped

½ teaspoon ground black pepper

1½ pounds medium shrimp, peeled and deveined

Salt to taste

2 green onions, thinly sliced

**DIRECTIONS**

1  In plastic bag, combine marinade ingredients (everything but the shrimp, salt, and onions). Remove ½ cup marinade; set aside for dipping sauce.

2  Add shrimp to marinade in bag, turning to coat. Close bag securely and marinate in refrigerator 30 minutes or up to 12 hours.

3  Remove shrimp from marinade; discard marinade.

4  Thread shrimp onto 8 skewers, dividing evenly.

5  Grill over medium coals 4 to 6 minutes or until shrimp turn pink and are just firm to the touch, turning once. Season with salt, as desired.

6  Meanwhile, prepare dipping sauce by placing reserved ½ cup marinade in a small saucepan. Bring to a boil over medium-high heat. Boil 3 to 5 minutes or until slightly reduced. Stir in green onions.

**PER SERVING (1/4 OF RECIPE):** *Calories: 202 (From Fat 13%); Fat: 3g; Cholesterol: 259mg; Sodium: 511mg; Carbohydrate: 7g (Dietary Fiber >1g); Protein: 35g.*

# Broiled Scallops with Honey-Lime Marinade

**PREP TIME: 15 MIN | COOK TIME: 7 MIN | YIELD: 2 SERVINGS**

### INGREDIENTS

2 tablespoons honey

1 tablespoon vegetable oil

4 teaspoons lime juice

¼ teaspoon grated lime peel

Hot pepper sauce

¼ teaspoon salt

½ pound bay, calico, or sea scallops

1 lime, cut in wedges

### DIRECTIONS

1 Combine honey, oil, lime juice, lime peel, hot pepper sauce, and salt. Combine all ingredients in a small bowl and whisk briskly until mixed well.

2 Pat scallops dry with paper towel and add to marinade. Marinate, stirring occasionally, up to 1 hour or cover and refrigerate, stirring occasionally, up to 24 hours.

3 Preheat broiler. Arrange scallops and marinade in a single layer in 2 individual broiler-proof dishes or in scallop shells.

4 Broil 4 inches from source of heat 4 to 7 minutes, depending on size of scallops, or until opaque throughout and lightly browned.

5 Serve with lime wedges and, if desired, hot, crusty French bread to soak up the juices. Can be doubled, tripled, or halved.

**PER SERVING (1/2 OF RECIPE):** *Calories: 240 (From Fat 29%); Fat: 8g; Cholesterol: 48mg; Sodium: 514mg; Carbohydrate: 23g (Dietary Fiber <1g); Protein: 21g.*

# A Honey of a Chili

**PREP TIME: 15 MIN | COOK TIME: 20 MIN | YIELD: 8 SERVINGS**

### INGREDIENTS

One 15-ounce package firm tofu

1 tablespoon vegetable oil

1 cup chopped onion

¾ cup chopped green bell pepper

2 cloves garlic, finely chopped

2 tablespoons chili powder

1 teaspoon ground cumin

1 teaspoon salt

½ teaspoon dried oregano

½ teaspoon crushed red pepper flakes

One 28-ounce can diced tomatoes, undrained

One 15½-ounce can red kidney beans, undrained

One 8-ounce can tomato sauce

¼ cup honey

2 tablespoons red wine vinegar

### DIRECTIONS

1 Using a cheese grater, shred tofu and freeze in a zippered bag or an airtight container.

2 Thaw tofu; place in a strainer and press out excess liquid.

3 In large saucepan or Dutch oven, heat oil over medium–high heat until hot; cook and stir onion, green pepper, and garlic 3 to 5 minutes or until vegetables are tender and begin to brown.

4 Stir in chili powder, cumin, salt, oregano, and crushed red pepper.

5 Stir in tofu; cook and stir 1 minute.

6 Stir in diced tomatoes, kidney beans, tomato sauce, honey, and vinegar.

7 Bring to a boil; reduce heat and simmer, uncovered, 15 to 20 minutes, stirring occasionally.

**PER SERVING:** *Calories: 295 (From Fat 26%); Fat: 9g; Cholesterol: 0mg; Sodium: 1,022mg; Carbohydrate: 41g (Dietary Fiber 9g); Protein: 18g.*

# Beef and Potato Tzimmes

**PREP TIME: 25 MIN | COOK TIME: 120 MIN | YIELD: 6 SERVINGS**

### INGREDIENTS

2 tablespoons vegetable oil, divided

2 pounds stew meat, cut in 1½-inch chunks

2 cups chopped onion

2 cups sliced (1-inch thick) carrots

2 teaspoons garlic salt

Water

2 cups cubed (1-inch thick) potato

2 cups cubed (1-inch thick) sweet potato

⅓ cup honey

½ teaspoon ground cinnamon

⅛ teaspoon ground pepper

4 ounces dried apricots

4 ounces pitted prunes

2 tablespoons flour, optional

2 tablespoons chopped parsley

### DIRECTIONS

1 Heat 1 tablespoon of oil in a heavy, 5-quart pot over medium heat. Add beef and brown on all sides.

2 Remove beef from pan, add remaining oil, if necessary, and sauté onion until tender.

3 Return beef to pan; add carrots, salt, and about 4 cups water to cover ingredients. Bring to a boil, reduce heat, cover, and simmer 1 hour.

4 Add potatoes, sweet potatoes, honey, cinnamon, and pepper; stir and return to a boil. Reduce heat and simmer, partially covered, 30 minutes or until potatoes are barely cooked.

5 Add dried fruit and simmer, uncovered, 30 minutes or until beef is tender. Liquid should be slightly thickened. If necessary, dissolve flour in 3 tablespoons water and stir into stew; return to simmer, stirring frequently.

6 Sprinkle with parsley before serving, if desired.

*PER SERVING: Calories: 616 (From Fat 23%); Fat: 15.9g; Cholesterol: 111mg; Sodium: 716mg; Carbohydrate: 79.7g (Dietary Fiber 7.8g); Protein: 41.6g.*

# Chewy Honey Oatmeal Cookies

**PREP TIME: 15 MIN | COOK TIME: 14 MIN | YIELD: 24 COOKIES**

**INGREDIENTS**

½ cup butter or margarine, softened

½ cup granulated sugar

½ cup honey

1 large egg

1 teaspoon vanilla extract

1½ cups quick-cooking rolled oats

1 cup whole-wheat flour

¼ teaspoon salt

1 teaspoon ground cinnamon

½ teaspoon baking soda

1 cup raisins, or chocolate or butterscotch chips

**DIRECTIONS**

1 Preheat oven to 350 degrees.

2 In medium bowl, beat butter with sugar until thoroughly blended.

3 Blend in honey. Blend in egg and vanilla, mixing until smooth.

4 In separate bowl, mix together oats, flour, salt, cinnamon, and baking soda; blend into honey mixture. Blend in raisins or chips.

5 Drop dough by rounded tablespoons onto a greased baking sheet. Bake for 12 to 14 minutes until cookies are golden brown.

6 Remove from oven and allow cookies to cool 2 to 3 minutes before removing from baking sheet. Cool completely, then store in an airtight container.

**PER SERVING (1 COOKIE WITH RAISINS):** *Calories: 130 (From Fat 29%); Fat: 4.5g; Cholesterol: 20mg; Sodium: 85mg; Carbohydrate: 23g (Dietary Fiber 2g); Protein: 2g.*

# Apple Honey Tart

**PREP TIME: 30 MIN** | **COOK TIME: 25 MIN** | **YIELD: 4 SERVINGS**

## INGREDIENTS

One 17¼-ounce puff pastry dough

1 egg, well beaten

1 cup white zinfandel wine*

½ cup honey

One 3-inch stick cinnamon

3 whole cloves

One ¼-inch slice fresh ginger root

3 medium apples, pared, cored, and sliced

Whipped cream or low-fat dairy sour cream

*If desired, apple juice may be substituted for wine.

## DIRECTIONS

1 Cut two 5-inch hearts out of puff pastry. Cut ½-inch-wide strips of pastry from remaining dough.

2 Brush edges of hearts with beaten egg.

3 Twist and line edges of hearts with dough strips, joining ends of strips with egg mixture as necessary. Bake according to package directions.

4 When golden and baked, remove or push down puffy centers of hearts to allow space for apple filling.

5 Bring wine, honey, and spices to boil in 9- to 10-inch skillet; reduce heat, cover, and simmer 10 to 15 minutes.

6 Add apples in one layer, return mixture to boil, and simmer 10 to 15 minutes or until apples are tender.

7 Carefully remove slices from liquid and drain thoroughly. Reduce liquid until syrupy; cool.

8 Brush bottom of crust with syrup; arrange poached apples over syrup.

9 Serve with dollops of whipped or sour cream.

**PER SERVING:** *Calories: 837 (From Fat 51%); Fat: 45g; Cholesterol: 53mg; Sodium: 525mg; Carbohydrate: 94.3g (Dietary Fiber 3.5g); Protein: 6.4g.*

# 7

# Appendixes

Find out where you can find all kinds of supplies and resources for your bees and hives, from medications to periodicals.

Use the Beekeeper's Checklist to ensure you're covering all of your bee bases at specific times of the year.

Check out the glossary for comprehensive definitions of all the words and terms you need to know.

# Appendix A

# Helpful Resources

As a new beekeeper, you'll welcome all the information that you can get your hands on. In this chapter, I present a bunch of resources that I find mighty useful: websites, vendors, associations, and journals.

## Honey Bee Information Websites

What in the world did I — or anyone — do before cyberspace? For one thing, I took a lot of trips to the library. But not even the most determined library search of years gone by would have turned up the plethora of bee-related resources that are only a click away on the web. Just enter the word "beekeeping" or "honey bees" into any of the search engines, and you'll come up with hundreds (even thousands) of finds. Like all things on the Net, many of these sites tend to come and go. A few are outstandingly helpful. Some are duds. Others have ridiculous information that may lead you to trouble. Each of the following sites is worth a visit.

### Apiservices — Virtual beekeeping gallery

www.apiservices.com

This European site is a useful gateway to scores of other beekeeping sites: forums, organizations, journals, vendors, conferences, images, articles, catalogs, apitherapy, beekeeping software, plus much more. It can be accessed in English, French, Spanish, and German and is nicely organized.

### The Barefoot Beekeeper

www.biobees.com

Phil Chandler offers lots of information on natural beekeeping and Top Bar hives, including a full set of plans and instructions on how to build your own Kenyan Top Bar hive.

# BeeHoo — The beekeeping directory

www.beehoo.com

This comprehensive international beekeeping directory has many helpful articles, information sheets, instructional guides, resources, photos, and links of interest for the backyard beekeeper. The site is viewable in English or in French and is definitely worthy of a bookmark.

# Beemaster Forum

www.beemaster.com

A popular international beekeeping forum designed to entertain and educate anyone with an interest in bees or beekeeping. Here you can share images, send messages, and participate in live forums. This secure site is moderated and is completely family friendly. It was created and is maintained by hobbyist beekeeper John Clayton (who also provided images in this book).

# Bee-Source.com

www.beesource.com

This site includes a nicely organized collection of bee-related articles, resources, and links, and it features sections on bees in the news, editorials, an online bookstore, a listing of beekeeping suppliers, plans for building your own equipment, discussion groups, bulletin boards, and much more.

# Facebook — Top Bar Beekeeping

Top Bar Beekeeping is a closed group on Facebook that has more than 3,200 members with more joining all the time. There are some lively discussions that can provide you with basic information and help you start thinking more about a subject. As with all Facebook pages, the information is mostly anecdotal, so checking things out is a must before you adapt any techniques being promoted.

# Mid-Atlantic Apiculture Research and Extension Consortium (MAARAC)

agdev.anr.udel.edu/maarec/

This research and extension consortium is packed with meaningful information for beekeepers worldwide. Download extension publications; find out more about videos, slide shows, software, and courses that are available from the organization; and read about honey-bee research currently underway. You can also discover important local beekeeping events planned in the Mid-Atlantic region and other national and international meetings of importance to beekeepers.

## National Honey Board

www.honey.com

This nonprofit government agency supports the commercial beekeeping industry. The folks at NHB are enormously helpful and accommodating. The well-designed site is a great source for all kinds of information about honey. You'll find articles, facts, honey recipes, and plenty of beautiful images.

# Bee Organizations and Conferences

Here are my favorite national and international beekeeping associations. Joining one or two of these is a great idea because their newsletters alone are worth the price of membership (dues are usually modest). Most of these organizations sponsor meetings and conferences. Attending one of these meetings is a fantastic way to learn about new tricks, find new equipment, and meet some mighty nice people with similar interests.

## American Apitherapy Society

This nonprofit organization researches and promotes the benefits of using honey-bee products for medical use. A journal is published by the society four times a year. Once a year, AAS organizes a certification course.

Phone 631-470-9446
Website www.apitherapy.org

## American Beekeeping Federation

This nonprofit organization plays host to a large beekeeping conference and trade show each year. The meetings are worth attending because they include a plethora of interesting presentations on honey bees and beekeeping. By all means join this organization to take advantage of its bimonthly newsletter. The organization's primary missions are benefiting commercial beekeepers and promoting the benefits of beekeeping to the general public.

3525 Piedmont Rd.
Bldg. 5, Ste. 300
Atlanta, GA 30305
Phone 404-760-2875
Email info@abfnet.org.
Website www.abfnet.org

## American Honey Producers

American Honey Producers is a nonprofit organization dedicated to promoting the common interest and general welfare of the American honey producer. The handsome website provides the public and other fellow beekeepers with industry news, membership information, convention schedules, cooking tips, and other information.

P.O. Box 435
Mendon, UT 84325
Phone 281-900-9740
Email cassie@AHPAnet.com
Website www.americanhoneyproducers.org

## Apiary Inspectors of America

The Apiary Inspectors of America is a nonprofit organization established to promote better beekeeping conditions in North America. Members of the Association, consisting of state apiarists, business representatives, and individual beekeepers, work collectively to establish more uniform and effective laws and methods for the suppression of honey-bee diseases, as well as a mutual understanding and cooperation between apiary inspection officials.

Website www.apiaryinspectors.org/

## Apimondia: International Federation of Beekeepers' Associations

Apimondia is a huge international organization composed of national beekeeping associations from all over the world, representing more than 5 million members.

The organization plays host to a large international conference and trade shows every other year.

Corso Vittorio Emanuele 101
I-00186 Roma
Italy
Phone +39 066852286
Fax +39 066852287
Website www.apimondia.org

## Eastern Apiculture Society

The Eastern Apiculture Society (EAS) was established in 1955 to promote honey-bee culture, the education of beekeepers, and excellence in bee research. Membership consists mostly of beekeepers east of the Mississippi River. Every summer, EAS conducts its annual conference in one of its 22 member states/provinces. The event is simply wonderful for a beekeeper. You can even take a comprehensive exam to become certified as an EAS "master beekeeper." By all means, try to attend one of these weeklong adventures.

Website www.easternapiculture.org

## International Bee Research Association

Founded in 1949, the International Bee Research Association (IBRA) is a nonprofit organization with members in almost every country in the world. Its mission is to increase awareness of the vital role of bees in agriculture and the natural environment. The organization is based in the United Kingdom. IBRA publishes several journals and sponsors international beekeeping conferences.

91 Brinsea Rd.,
Congresbury,
Bristol,
BS49 5JJ
United Kingdom
Phone 00 44 (0) 29 2037 2409
Email mail@ibra.org.uk
Website www.ibrabee.org.uk

## USDA Agricultural Research Service

Known as Beltsville in the bee world, the Bee Research Laboratory is a division of the U.S. Department of Agriculture and is a good agency to know about. After all, if you're an American, your tax dollars are paying for it! The mission of the Bee Research Laboratory in Beltsville is to conduct research on the biology and control of honey-bee parasites, diseases, and pests to ensure an adequate supply of bees for pollination and honey production. The list of scientists who have worked at Beltsville in the past reads like a "Who's Who of American Beekeeping Research." If you ever need to (let's hope not), you can send samples of your sick bees to the lab for analysis. The lab also is consulted when there's a question about whether a colony is Africanized.

Bee Research Laboratory
10300 Baltimore Ave.
Bldg. 306, Room 315, BARC-East
Beltsville, MD 20705
Phone 301-504-5143
Website ars.usda.gov/main/site_main.htm?modecode=12-45-33-00

## The Western Apiculture Society

The Western Apicultural Society is a nonprofit, educational, beekeeping organization founded in 1978 for the benefit and enjoyment of all beekeepers in western North America.

Website www.westernapiculturalsociety.org

# Bee Journals and Magazines

Are you ready to curl up with a good article about honey bees? A bunch of publications are worth a read. Subscribing to one or more of them provides you with ongoing sources of useful beekeeping tips and practical information. And the ads in these journals are a great way to learn about new beekeeping toys and gadgets. Here are some English-language journals of interest.

## American Bee Journal

The *American Bee Journal* was established in 1861 and has been published continuously since that time, except for a brief period during the Civil War. The *Journal* has the honor of being the oldest English-language beekeeping publication in the world.

Today, Dadant and Sons publishes the *American Bee Journal* for subscribers through-out the world. Readership is concentrated among hobby and commercial beekeep-ers, bee supply dealers, queen breeders, shippers, honey packers, and entomologists.

Dadant & Sons, Inc.
51 S. Second St.
Hamilton, IL 62341
Phone 217-847-3324
Fax 217-847-3324
Email info@americanbeejournal.com
Website americanbeejournal.com

## Bee Culture

*Bee Culture* has been around since the late 1800s. Articles are aimed at the needs and interests of the backyard beekeeper and small-scale honey producers. This journal features a wide range of "how-to" articles, Q&A, honey recipes, and industry news. This is the "bible" for the hobbyist beekeeper.

The *Bee Culture* website lists "Who's Who in North American Beekeeping." This terrific database enables you to search for bee clubs in your area. Go to www.beeculture.com/?s=who%27s+who.

A.I. Root Company
623 W. Liberty St.
Medina, OH 44256
Phone 800-289-7668, Ext. 3220
Website www.beeculture.com

**TIP**

Be sure to check out the coupon later in this book that gives you a discount on a subscription to *Bee Culture*.

## Bee World

*Bee World*, founded in 1919, is a journal that digests research studies and articles from around the world. Reliable and practical information comes from bee experts worldwide. It's published by the International Bee Research Association (IBRA).

International Bee Research Association
91 Brinsea Rd.,
Congresbury,
Bristol,

BS49 5JJ
United Kingdom
Phone 00 44 (0) 29 2037 2409
Email mail@ibra.org.uk
Website www.ibrabee.org.uk/index.php/ibra-bee-research-journals-publications/item/3119

# Beekeeping Supplies and Equipment

Where do you find all the neat stuff you need to become a beekeeper? Where do you buy bees? Start with a search on the web for "beekeeping supplies." Maybe you'll get lucky and find a listing for a local supplier who sells supplies out of his or her garage. That's kind of cool because it gives you face-to-face access to your own personal mentor. Alternatively, you can deal directly with one of the major bee suppliers. They all offer mail-order catalogs, and all now have e-commerce-enabled websites. Some provide online advice (your online mentor). I've included many of the more popular suppliers in the United States and a couple of international suppliers.

## Bee-commerce.com

This online company based in Connecticut offers secure e-commerce shopping and personalized support designed exclusively for the backyard beekeeper. A free download section is provided with helpful instruction sheets and articles. The company's beekeeping experts happily serve as your online mentors.

Hopewell Harmony, LLC
160A Sugar St.
Newtown, CT 06470
Phone 800-784-1911 or 203-222-2268
Email info@bee-commerce.com
Website www.bee-commerce.com

**TIP**

See the offer at the end of this book for a special coupon from this vendor.

## Bee Thinking

Bee Thinking has a retail shop in Portland, Oregon, plus a nice website where you can shop for beekeeping equipment, supplies, books, gifts, and even live bees.

There are also helpful links to useful resources. Bee Thinking manufactures beautiful Western red-cedar Langstroth, Top Bar, and Warré hives. At Bee Thinking, the emphasis is on "natural" beekeeping practices.

Bee Thinking, LLC
1551 SE Poplar Ave.
Portland, OR 97214
Phone 877-325-2221
Email support@beethinking.com
Website www.beethinking.com

See the offer at the end of this book for a special coupon from this vendor.

## BeeWeaver Apiaries

Established in 1888, the company provides live bees and hybrid queens across the country and worldwide. Remember to order/reserve your live bees well before the spring that you need them. Apiaries can sellout quickly. Alas, the only way to reach the company seems to be via a form on its website. It offers no email address and publishes no phone number.

BeeWeaver Apiaries
Brazos Valley Location
16481 CR 319
Navasota, TX 77868

Texas Bee Barn
Hill Country Location
3700 McGregor Lane
Dripping Springs, TX 78620
Website www.beeweaver.com

See the offer at the end of this book for a special coupon from this vendor.

## Betterbee

Betterbee offers online and mail-order shopping for beekeeping supplies and equipment. Betterbee also has a great selection of supplies for candle making, soap making, and even a learning center with relevant articles and how-to information. It also offers local pickup of live bee packages. Worth a visit.

8 Meader Rd.
Greenwich, NY 12834
Phone 800-632-3379 or 518-314-0575

Email support@betterbee.com
Website www.betterbee.com

## Brushy Mountain Bee Farm

Brushy Mountain Bee Farm features a large selection of beekeeping supplies and equipment by mail order and e-commerce. Brushy Mountain manufactures quality hives and offers a wide range of top-quality equipment. It has brick-and-mortar stores in North Carolina, Pennsylvania, and Oregon.

610 Bethany Church Rd.
Moravian Falls, NC 28654
Phone 800-233-7929
Email sales@brushymountainbeefarm.com
Website www.brushymountainbeefarm.com

**TIP**

See the offer at the end of this book for a special coupon from this vendor.

## Dadant & Sons, Inc.

Dadant & Sons, Inc. provides beekeeping supplies and equipment by mail order and e-commerce. It has several regional offices around the United States. One of the largest beekeeping suppliers in the United States, it has been in business since 1878.

51 South Second St.
Hamilton, IL 62341
Phone 888-922-1293 or 217-847-3324
Fax 217-847-3660
Email dadant@dadant.com
Website www.dadant.com

**TIP**

See the offer at the end of this book for a special coupon from this vendor.

## Glorybee Foods, Inc.

Beekeeping equipment plus a wide array of soap, skin care, aromatherapy, and candle-making supplies are available by mail order and e-commerce from Glorybee Foods, Inc. It also is a great resource for bee-related gifts.

29548 B Airport Rd.
P.O. Box 2744
Eugene, OR 97402
Phone 800-456-7923
Email info@glorybeefoods.com
Website www.glorybeefoods.com

See the offer at the end of this book for a special coupon from this vendor.

# BeeInventive

BeeInventive created and markets the Flow hive. Based in Australia, the company ships this hive around the world through distributors in Europe and the United States. The hive can only be purchased using the company's website, which offers lots of information and videos on the Flow hive.

BeeInventive Pty Ltd
ABN 31 161 952 941
Phone 646-876-8880 (U.S.); +61 2 8880 0774 (Australia);
+3 18 5208 4111 (Europe [Netherlands])
Website www.honeyflow.com

See the offer at the end of this book for a special coupon from this vendor.

# Kelley Beekeeping

Kelley Beekeeping provides beekeeping supplies and equipment by mail order and e-commerce. It offers a selection of quality products.

807 W. Main St.
Clarkson, KY 42726
Phone 800-233-2899 or 270-242-2012
Fax 270-242-4801
Email kelleybees@kynet.net
Website www.kelleybees.com

See the offer at the end of this book for a special coupon from this vendor.

# Mann Lake

Mann Lake, Ltd. offers beekeeping supplies, equipment, and medication by mail order and e-commerce. Its website features an online catalog and a very helpful beekeeping learning center. It has store locations in Minnesota, Pennsylvania, and California.

501 First St. S.
Hackensack, MN 56452
Phone 800-880-7694
Fax 218-675-6156
Email beekeeper@mannlakeltd.com
Website www.mannlakeltd.com

See the offer at the end of this book for a special coupon from this vendor.

# Miller Bee Supply

A nice, easy-to-use site offering a wide assortment of start-up kits, beekeeping supplies, medications, and queen-rearing equipment.

496 Yellow Banks Rd.
North Wilkesboro, NC 28659
Phone 888-848-5184 or 336-670-2249
Fax 336-670-2293
Email Info@Millerbeesupply.com
Website www.millerbeesupply.com

See the offer at the end of this book for a special coupon from this vendor.

# Rossman Apiaries

Rossman Apiaries, located in South Georgia and serving beekeepers since 1987, carries a wide range of supplies and manufactures highly durable Cypress Woodenware Hives. Rossman also raises and sells packages of Italian bees and queens.

3364-A GA Hwy. 33 N.
P.O. Box 909
Moultrie, GA 31776-0909
Phone 800-333-7677
Email rossmanbees@windstream.net
Website www.gabees.com

See the offer at the end of this book for a special coupon from this vendor.

## Sacramento Beekeeping

A friendly retail store and online site offering beekeeping supplies, soap- and candle-making supplies, gifts, and honey. Family-owned and operated.

2110 X St.
Sacramento, CA 95818
Phone 916-451-2337
Fax 916-451-7008
Email info@sacramentobeekeping.com
Website www.sacramentobeekeeping.com

## Swienty Beekeeping Equipment

European supplier Swienty Beekeeping Equipment offers beekeeping supplies by mail order and e-commerce, including a nice selection of unique products not readily available in the United States. Its website is published in three languages. *Note:* It is a source for a very nice honey press mentioned in Chapter 16.

Hørtoftvej 16
6400 Sønderborg
Denmark
Phone +45 7448 6969
Fax +45 7448 8001
Email shop@swienty.com
Website www.swienty.com

See the offer at the end of this book for a special coupon from this vendor.

## Thorne Beekeeping Supply

Thorne offers a wide range of beekeeping supplies and equipment, including a great selection of different hive styles (such as National, Warré, and Top Bar hives). You will also find candle-making supplies and useful information to download, plus links to helpful beekeeping-related sites. A nicely designed site.

EH Thorne (Beehives) Ltd
Beehive Business Park
Rand
Nr Wragby

Market Rasen
LN8 5NJ
United Kingdom
Phone +44 (0)1 673 858 555
Email sales@thorne.co.uk
Website www.thorne.co.uk

# State Bee Inspectors (United States)

If you live in the United States, you will want to know how to contact the bee inspector in your state. The inspector is there to help. If you have a pesky problem with your bees' health or a question needing the attention of a bee expert, call your state bee inspector. Alas, with governmental budget cuts, not all regions have a designated bee inspector. To track them down, just go to this web address: www.beeculture.com/apiary-inspectors/.

# Beekeeper's Checklist

Use this simple checklist every time you inspect your bees. It's okay to photocopy it. Be sure to date copies and keep them in a binder for future reference and comparison. Use one form for each of your hives. For more details on what to look for during inspections, be sure to read Chapters 8 through 13. Happy beekeeping!

**Hive number/location** _____

**Date of this inspection** _____

**Date queen/hive was established** _____

| Observations | Notations |
|---|---|
| ❑ Observe bees on the ground in front of the hive. Do they appear to be staggering or crawling up grass blades and then falling off? If yes, this may be an indication of a virus or tracheal mites. Take steps accordingly. | |
| ❑ Observe bees at entrance. (Look for dead bees or abnormal behavior and appearance.) | |
| ❑ Do you see significant "spotting" of feces on the hive? (If yes, the bees may have Nosema and need to be treated.) | |
| ❑ What is the condition of your equipment? (Note any needed repairs that have to be made or replacement parts to order.) | |
| ❑ How's the brood pattern? (It should be compact and plentiful during the brood-rearing season.) | |
| ❑ Check appearance of brood cappings. (Cappings should be slightly convex and free of perforations.) | |
| ❑ Evaluate your queen based on her egg-laying ability. (Do you need to replace her with a new queen?) | |
| ❑ How do the larvae look? (Larvae should be a glistening, snowy white.) | |
| ❑ Is the colony healthy? (You should find lots of active bees, healthy-looking brood, a clean hive, and a nice sweet smell.) | |
| ❑ Do you see eggs? (You should find only one egg per cell.) | |

*(continued)*

| Observations | Notations |
|---|---|
| ❏ Can you find the queen? (Is she the same "marked" one you introduced?) | |
| ❏ Check for swarm cells. (Take swarm prevention steps, if needed.) | |
| ❏ Check for supercedure cells. (May be an indication that your queen is underperforming and needs to be replaced.) | |
| ❏ Do you see evidence of Varroa mites (on bees or on sticky board)? If yes, take corrective action accordingly. | |
| ❏ Do the bees have food? (They need honey, pollen, and nectar.) | |
| ❏ How much capped honey is there? (Is it time to add a queen excluder and honey supers to your Langstroth hive?) | |
| ❏ Do the bees have an adequate water supply? | |
| ❏ Is it time to feed? (This usually is done in spring and autumn, depending on where you live.) | |
| ❏ Check ventilation. (Adjust based on weather conditions.) | |
| ❏ Clean off propolis and burr comb that make manipulation difficult. | |
| ❏ What did you do for hive manipulations? | |

## Action Items (What to do between now and the next inspection):

_____

_____

_____

_____

_____

_____

_____

_____

_____

_____

_____

_____

# Glossary

**abscond:** To leave a hive suddenly, usually because of problems with poor ventilation, too much heat, too much moisture, mites, moths, ants, beetles, lack of food, or other intolerable problems.

**acarine disease:** The name given to the problems bees experience when they are infested with tracheal mites *(Acarapis woodi)*.

**Africanized honey bee (AHB):** A short-tempered and aggressive bee that resulted from the introduction of African bees into Brazil, displacing earlier introductions of more gentle European bees. The media has dubbed Africanized bees as "killer bees" because of their aggressive behavior.

**apiary:** This is the specific location where a hive(s) is kept. (Sometimes referred to as a **bee yard**.)

**apiculture:** The science, study, and art of raising honey bees. (As a beekeeper, you are an apiculturist!)

***Apis mellifera*:** The scientific name for the European honey bee.

**apitherapy:** The art and science of using products of the honey bee for therapeutic/medical purposes.

**battery box:** Ventilated container used for shipping or housing caged queen bees and a small population of attending worker bees.

**bee bread:** Pollen, collected by bees, that is mixed with various liquids and then stored in cells for later use as a high-protein food for larvae and bees.

**bee space:** The critical measurement between parts of a hive that enables bees to move freely about the hive. The space measures $\frac{1}{4}$ to $\frac{3}{8}$ inch (7 to 9 millimeters).

**bee veil:** A mesh, see-through netting worn over the head to protect the beekeeper from stings.

**beehive:** The "house" where a colony (family) of honey bees lives. In nature, it may be the hollow of an old tree. For the beekeeper, it usually is a boxlike device containing frames of beeswax comb.

**beeswax:** The substance secreted by glands in the worker bee's abdomen that is used by the bees to build comb. It can be harvested by the beekeeper and used to make candles, cosmetics, and other beeswax products.

**bottom board:** The piece of the hive that makes the ground floor.

**brace/burr comb:** Brace comb refers to the bits of random comb that connect two frames, or any hive parts, together. Burr comb is any extension of comb beyond what the bees build within the frames. (Both should be removed by the beekeeper to facilitate manipulation and inspection of frames.)

**brood:** A term that refers to immature bees, in the various stages of development, before they have emerged from their cells (eggs, larvae, and pupae).

**brood chamber:** The part of the hive where the queen lays eggs and the brood is raised. This is typically the lower deep.

**capped brood:** The pupal cells that have been capped with a wax cover, enabling the larvae to spin a cocoon and turn into pupae.

**castes:** The two types of female bees (workers and queens).

**cell:** The hexagon-shaped compartment of a comb. Bees store food (honeycomb) and raise brood (brood comb) in these compartments (cells).

**cleansing flight:** Refers to when bees fly out of the hive to defecate after periods of confinement. (A good day to wear a hat.)

**cluster:** A mass of bees, such as a swarm. Also refers to when bees huddle together in cool weather.

**colony:** A collection of bees (worker bees, drones, and a queen) living together as a single social unit.

**colony collapse disorder (CCD):** Term given to the sudden die-off of honey bees in colonies.

**comb:** A back-to-back collection of hexagonal cells that are made of beeswax and used by the bees to store food (honeycomb) and raise brood (brood comb).

**crystallization:** The process by which honey granulates or becomes a solid (rather than a liquid).

**dancing:** A series of repeated bee movements that plays a role in communicating information about the location of food sources and new homes for the colony.

**deep hive body:** The box that holds standard full-depth frames. (A deep box is usually $9\frac{5}{8}$ inches deep. It is often simply referred to as a deep.)

**drawn comb:** A sheet of beeswax foundation (template) whose cells have been drawn out into comb using beeswax produced by the bees.

**drifting:** Refers to when bees lose their sense of direction and wander into neighboring hives. (Drifting usually occurs when hives are placed too close to each other.)

**drone:** The male honey bee whose main job is to fertilize the queen bee.

**egg:** The first stage of a bee's development (metamorphosis).

**entrance reducer:** A notched strip of wood placed at the hive's entrance to regulate the size of the "front door." Used mostly in colder months and on new colonies, it helps control temperatures and the flow of bees. It can also be used with new or small colonies to prevent robbing.

**extractor:** A machine that spins honeycomb and removes liquid honey via centrifugal force. (The resulting honey is called extracted honey or liquid honey.)

**feeder:** A device that is used to feed sugar syrup to honey bees.

**feral bees:** A wild honey bee colony that is not managed by a beekeeper.

**follower board:** Used in Top Bar hives, this snug-fitting device seals off a part of the hive chamber you don't want the bees to access. It can be moved to give the bees more or less space, depending upon the beekeeper's objectives.

**food chamber:** The part of the hive used by the bees to primarily store pollen and honey. This is typically the upper deep when two deep hive bodies are used.

**foulbrood:** Bacterial diseases of bee brood. American foulbrood is very contagious — it is one of the most serious bee diseases. European foulbrood is less threatening. Colonies can be treated with an antibiotic (such as Terramycin) to prevent or treat foulbrood.

**foundation:** A thin sheet of beeswax that has been embossed with a pattern of hexagon-shaped cells. Bees use this as a guide to neatly build full-depth cells into comb.

**frame:** Four pieces of wood that come together to form a rectangle designed to hold foundation that the bees draw into honeycomb.

**grafting:** The manual process of transporting young larvae into special wax or plastic "queen cell cups" in order to raise new queen bees.

**hive:** A home provided by the beekeeper for a colony of bees.

**hive tool:** A metal device used by beekeepers to open the hive and pry frames apart for inspection.

**honey flow:** The period of time when one or more major nectar sources (flowers) are in full bloom and the weather is perfect for the bees to fly and collect lots of nectar for making honey. (Sometimes called a **nectar flow**.)

**honey press:** A screw press tool used to squeeze and strain honey from the comb. It's used to harvest liquid honey from Top Bar hives.

**honeycomb:** Comb that has been filled with honey.

**inbreeding:** The production of offspring from the mating or breeding of bees that are closely related genetically, in contrast to outcrossing, which refers to mating unrelated individuals.

**inner cover:** A flat board with a ventilation hole that goes between the upper hive body and the outer (top) cover.

**larva (pl. *larvae*):** The second stage in the development (metamorphosis) of the bee.

**laying worker:** A worker bee that lays eggs. (Because they are unfertile, their eggs can only develop into drones.)

**marked queen:** A queen bee marked on the thorax with a dot of paint to make it easier to find her, document her age, or keep track of her.

**miticides:** Pesticide chemicals used to control mites.

**nectar:** The sweet, watery liquid secreted by plants. (Bees collect nectar and make it into honey.)

**nectar flow:** The period of time when one or more major nectar sources (flowers) are in full bloom and the weather is perfect for the bees to fly and collect lots of nectar. (Sometimes called a **honey flow**.)

**neonicotinoids:** A class of insecticides that act on the central nervous system of insects. Neonicotinoids are among the most widely used insecticides worldwide, but their use may have a connection to colony collapse disorder.

**Nosema disease:** An illness of the adult honey bee's digestive track caused by the fungal pathogens, Nosema apis and Nosema ceranae. The disease can be controlled with an antibiotic (such as Fumigilin-B).

**nucleus hive (nuc):** A small colony of bees housed in a three- to five-frame cardboard or wooden hive.

**nuptial mating flight:** The flight that a newly emerged virgin queen takes when she leaves the hive to mate with drones.

**nurse bees:** Young adult bees who feed the larvae.

**outcrossing:** The practice of introducing unrelated genetic material into a breeding line. It increases genetic diversity, thus reducing the probability of disease and/or reducing genetic abnormalities.

**outer cover:** The "lid" that goes on top of the hive to protect against the elements. (Sometimes called a **top cover** or **telescoping outer cover**.)

**pheromone:** A chemical scent released by an insect or other animal that stimulates a behavioral response in others of the same species.

**pollen:** The powdery substance that is the male reproductive cell of flowers. (Bees collect pollen as a protein food source.)

**propolis:** A sticky, resinous material that bees collect from trees and plants and use to seal up cracks and strengthen comb. It has antimicrobial qualities. (Also called **bee glue**.)

**pupa (pl. *pupae*):** The third and final stage in the honey bee's metamorphosis before it emerges from the capped cell as a mature bee.

**queen:** The mated female bee, with fully developed ovaries, that produces male and female offspring. (There is usually only one queen to a colony.)

**queen bank:** The storage of mated honey-bee queens in cages within a full-sized colony of bees.

**queen cage:** A small, screened box used to temporarily house queen bees (such as during shipment or when being introduced into a new hive).

**queen excluder:** A frame holding a precisely spaced metal grid. The device usually is placed immediately below the honey supers to restrict the queen from entering that area and laying eggs in the honeycomb. The spacing of the grid allows foraging bees to pass through freely, but it is too narrow for the larger-bodied queen to pass through.

**queen mothers:** Desirable queen(s) used as donors of eggs or young larvae to raise more queen bees.

**queen substance:** A term that refers to the pheromone secreted by the queen. It is passed throughout the colony by worker bees.

**queenless nuc:** A small colony of bees (without a queen) that is used to draw out introduced queen cells and nurture the queen larvae.

**reversing:** The managerial ritual of switching a colony's hive bodies to encourage better brood production. (Usually done in the early spring.)

**robbing:** Pilfering of honey from a weak colony by other bees or insects.

**royal jelly:** The substance that is secreted from glands in a worker bee's head and is used to feed the brood.

**scout bees:** The worker bees that look for pollen, nectar, or a new nesting site.

**shallow super:** The box that is used to collect surplus honey. The box is $5\,^{11}\!/_{16}$ inches deep. (Sometimes called a **honey super**.)

**smoker:** A tool with bellows and a fire chamber that is used by beekeepers to produce thick, cool smoke. The smoke makes colonies easier to work with during inspections.

**stinger:** The hypodermic-like stinger is located at the end of the adult female bee's abdomen. Remember, bees don't bite! They sting.

**supercedure:** The natural occurrence of a colony replacing an old or ailing queen with a new queen. (A cell containing a queen larva destined to replace the old queen is called a **supercedure cell**.)

**supering:** The act of adding (honey) supers to a colony.

**surplus honey:** Refers to the honey that is above and beyond what a colony needs for its own use. It is this "extra" honey that the beekeeper may harvest for his or her own use.

**swarm:** A collection of bees and a queen that has left one hive in search of a new home (usually because the original colony had become too crowded). Bees typically leave behind about half of the original colony and the makings for a new queen. The act itself is called swarming.

**top bars:** The wooden planks that are placed in Top Bar hives on which the bees build natural, free-form wax comb.

**uncapping knife:** A device used to slice the wax capping off honeycomb that is to be extracted. (These special knives usually are heated electrically or by steam.)

**winter cluster:** A tightly packed colony of bees, hunkered down for the cold winter months.

**worker bee:** The female honey bee that constitutes the majority of the colony's population. Worker bees do most of the chores for the colony (except egg laying, which is done by the queen).

# Index

## M

MAARAC (Mid Atlantic Apiculture Research and Extension Consortium), 400–401

magnifying goggles, 41

mail order, 85

making a divide, 354

*Making Mead Honey Wine* (Morse), 367

malva (*Malvaceae*), 364

*Malvaceae*, 364

*Malva sylvestris*, 364

mandibles, honey bee, 26

manna, 306

Mann Lake, 410

manuka honey, 310

marked queen bee, 125, 382, 418

mason bee, 45

mason-bee nest, 46

matches, 119

mated queens. *See also* queen bees; raising queens

  marking, 296

  queenless nuc, 283

  storing, 295

mating behavior, 38, 282

mating flights, 38, 282

Mayo Clinic, 302

mead, brewing, 365–367

medicated beekeeping, 20

medications

  autumn, administering, 187

  decisions, 242

  foulbrood, 195

  new, 244

  Nosema, 195

  spring, administering, 195–196

  tracheal mites, 196

  Varroa mites, 195

medium honey super, 66, 91–92

*Mentha*, 364

menthol crystals, 196, 262–263

mesquite honey, 310

Metheglin, 366

Mid Atlantic Apiculture Research and Extension Consortium (MAARAC), 400–401

migratory beekeepers, 13

Miller Bee Supply, 410

Miller feeders. *See* hive-top feeders

Miller method. *See also* queen rearing

  defined, 285

  process, 285–289

  queen cells, 286–287

  queen emergence, 288

  queenless nuc, 286–287

  wax foundation, 285–286

minerals, 301

mini-mating nucs, 293–294

mint (*Mentha*), 364

Mite-a-Thol (menthol crystals), 262–263

Mite Away Quick Strips (formic acid), 258, 264

mites

  absconding and, 217

  inspections, 383

  tracheal, 259–264

  Varroa mites, 252–259, 383

miticides, 258–259, 418

moisture, 317

molded candles, beeswax, 371

*Monarda citriodora*, 363

*Monarda didyma*, 363

*Monarda punctata*, 363

Morse, Roger (author), 367

  *Making Mead Honey Wine*

mountain honey, 311

mouse, 270–272, 273. *See also* pests

mouse guards

  installing, 271

  placing, 190

  use dividends, 271

mouth, honey bee, 26

must (honey and water mixture), 366

## N

Nasonov gland, 35

nasturtium (*Tropaeolum minus*), 364

National Honey Board, 385, 401

natural beekeeping, 20

*Natural Beekeeping* (Conrad), 20

natural comb, 173

natural honey, labels, 330

nectar, 418

nectar flow, 419

neighbors, easing minds of, 55–56, 59

# About the Author

**Howland Blackiston** has been a backyard beekeeper since 1984. He's written many articles on beekeeping and appeared on dozens of television and radio programs (including The Discovery Channel, CNBC, CNN, NPR, Sirius Satellite Radio, and scores of regional shows). He has been a featured speaker at conferences in more than 40 countries. Howland is the past president of Connecticut's Back Yard Bee-keepers Association, one of the nation's largest regional clubs for the hobbyist beekeeper. He is also the author of *Building Beehives For Dummies,* the companion book to this one, providing detailed instructions in how to build hives from scratch. Howland, his wife, Joy, and his bees live in Easton, Connecticut.

# Dedication

This book is lovingly dedicated to my wife, Joy, who is the queen bee of my universe. She has always been supportive of my unconventional whims and hobbies (and there are a lot of them), and never once did she make me feel like a dummy for asking her to share our lives with honey bees. I also thank our wonderful daughter, Brooke (now grown, married, and a mom), who, like her mother, cheerfully put up with sticky kitchen floors and millions of buzzing "siblings" while growing up in our bee-friendly household.

# Author's Acknowledgments

I was very fortunate, when I started beekeeping, that I met a masterful beekeeper who took me under his wing and taught me all that is wonderful about honey bees. Sadly, Ed Weiss passed away in 2016, but over the years he became a valued mentor and a great friend. Thank you, Ed, for getting me started in this wonderful hobby.

My good friends Anne Mount and David Mayer played a key role in the creation of this book. Both of them are authors, and both encouraged me to contact the *For Dummies* team at Wiley Publishing. "You should write a book about beekeeping, and they should publish it," they urged. Well, I did, and they did. Thank you, Anne and David. I owe you a whacking big jar of honey!

A good how-to book needs lots of great how-to images. Many thanks to the following who provided the images in this edition: William Hesbach, John Clayton, Edward Ross, Eric Erickson, Mario Espinola, Michael Joshin Thiele, Amanda Lane, Kate Solomon, Marina Marchese, Lynda Richardson, Sam Droege, Sharon Stiteler, Dave Stobbe, Virginia Williams, Rob Snyder, Jeff Pettis, Alex Wild, Scott Bauer, Stephen Ausmus, William Styer, Daniel Caron, Dave Stobbe, Miranda Sherman, and Kim Flottum at *Bee Culture* magazine. Additional image credits also go to Bee-Inventive Pty Ltd, The Informed Partnership, The National Honey Board, the U.S. Department of Agriculture, U.S. Geological Survey, Bee Thinking, Brushy Mountain Bee Farm, Kelley Beekeeping, Bee-Commerce, BeeSmart, HiveTracks, Glorybee, Misco Refractometers, Swienty Beekeeping Equipment, Wilbanks Apiaries, Mann Lake Bee Supplies, and The National Honey Board. And thanks to fellow beekeeper and friend Stephan Grozinger, who patiently served as my model for some of the how-to photographs.

I started planning this fourth edition by organizing a small focus group of beekeepers to identify new topics, features, and content that would bring the greatest value to the book's readers. My thanks to backyard beekeepers Kathy Hammell, Paul Shelley, Denise Valentine, Rick Glover, and Jean Japinga for participating in this focus group.

This fourth edition contains many helpful bits of advice for you urban beekeepers who face your own set of unique challenges raising bees in a city environment. These words of urban wisdom come from my friend Andrew Cote, whom I regard as the wizard of urban beekeeping. Andrew manages 60 hives in New York City and has helped many metropolitan beekeepers get started in this wonderful hobby. Thank you, Andrew, for sharing your special know-how.

I have known and admired Dr. Dewey Caron for more than 20 years, and I was thrilled when he agreed to write the foreword and serve as the technical editor for this edition. As one of the most prominent entomologists in the United States, there is no one more qualified to ensure that the information and advice in the book is spot on. He was also a tremendous help including information on natural, chemical-free options for keeping your bees healthy. Thank you, Dewey.

I regard William Hesbach as an expert in the management of Top Bar hives, so I was thrilled when he agreed to help me develop new content for this edition that would be of particular interest to beekeepers wanting to keep bees in Top Bar hives.

Thanks also to my friends Leslie Huston for her help creating the chapter on raising queen bees, Marina Marchese for her help with the chapter about honey, Ellen Zampino for her contributions to the section on planting flowers for your bees, and Patty Pulliam for her wonderful beeswax recipes.

Generous thanks to the companies that provided tempting discounts and special offers via the coupons that appear at the end of this book. These suppliers added great additional value to the book.

Writing this book was a labor of love, thanks to the wonderful folks at Wiley Publishing: Vicki Adang, my development editor, who kept the project buzzing along; Tracy Boggier, my senior acquisitions editor; and Christy Pingleton, the book's copy editor, who made my written words sound very spiffy. What a great team!

## Publisher's Acknowledgments

**Senior Acquisitions Editor:** Tracy Boggier

**Development Editor:** Victoria M. Adang

**Copy Editor:** Christine Pingleton

**Technical Editor:** Dewey M. Caron

**Production Editor:** Antony Sami

**Cover Photos:** © StudioSmart/Shutterstock

## Apple & Mac

iPad For Dummies,
6th Edition
978-1-118-72306-7

iPhone For Dummies,
7th Edition
978-1-118-69083-3

Macs All-in-One
For Dummies, 4th Edition
978-1-118-82210-4

OS X Mavericks
For Dummies
978-1-118-69188-5

## Blogging & Social Media

Facebook For Dummies,
5th Edition
978-1-118-63312-0

Social Media Engagement
For Dummies
978-1-118-53019-1

WordPress For Dummies,
6th Edition
978-1-118-79161-5

## Business

Stock Investing
For Dummies, 4th Edition
978-1-118-37678-2

Investing For Dummies,
6th Edition
978-0-470-90545-6

Personal Finance
For Dummies, 7th Edition
978-1-118-11785-9

QuickBooks 2014
For Dummies
978-1-118-72005-9

Small Business Marketing
Kit For Dummies,
3rd Edition
978-1-118-31183-7

## Careers

Job Interviews
For Dummies, 4th Edition
978-1-118-11290-8

Job Searching with Social
Media For Dummies,
2nd Edition
978-1-118-67856-5

Personal Branding
For Dummies
978-1-118-11792-7

Resumes For Dummies,
6th Edition
978-0-470-87361-8

Starting an Etsy Business
For Dummies, 2nd Edition
978-1-118-59024-9

## Diet & Nutrition

Belly Fat Diet For Dummies
978-1-118-34585-6

Mediterranean Diet
For Dummies
978-1-118-71525-3

Nutrition For Dummies,
5th Edition
978-0-470-93231-5

## Digital Photography

Digital SLR Photography
All-in-One For Dummies,
2nd Edition
978-1-118-59082-9

Digital SLR Video &
Filmmaking For Dummies
978-1-118-36598-4

Photoshop Elements 12
For Dummies
978-1-118-72714-0

## Gardening

Herb Gardening
For Dummies, 2nd Edition
978-0-470-61778-6

Gardening with Free-Range
Chickens For Dummies
978-1-118-54754-0

## Health

Boosting Your Immunity
For Dummies
978-1-118-40200-9

Diabetes For Dummies,
4th Edition
978-1-118-29447-5

Living Paleo For Dummies
978-1-118-29405-5

## Big Data

Big Data For Dummies
978-1-118-50422-2

Data Visualization
For Dummies
978-1-118-50289-1

Hadoop For Dummies
978-1-118-60755-8

## Language &
## Foreign Language

500 Spanish Verbs
For Dummies
978-1-118-02382-2

English Grammar
For Dummies, 2nd Edition
978-0-470-54664-2

French All-in-One
For Dummies
978-1-118-22815-9

German Essentials
For Dummies
978-1-118-18422-6

Italian For Dummies,
2nd Edition
978-1-118-00465-4

**e Available in print and e-book formats.**

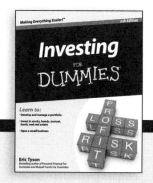
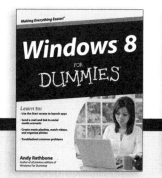

Available wherever books are sold. **For more information or to order direct visit www.dummies.com**

## Math & Science

Algebra I For Dummies,
2nd Edition
978-0-470-55964-2

Anatomy and Physiology
For Dummies, 2nd Edition
978-0-470-92326-9

Astronomy For Dummies,
3rd Edition
978-1-118-37697-3

Biology For Dummies,
2nd Edition
978-0-470-59875-7

Chemistry For Dummies,
2nd Edition
978-1-118-00730-3

1001 Algebra II Practice
Problems For Dummies
978-1-118-44662-1

## Microsoft Office

Excel 2013 For Dummies
978-1-118-51012-4

Office 2013 All-in-One
For Dummies
978-1-118-51636-2

PowerPoint 2013
For Dummies
978-1-118-50253-2

Word 2013 For Dummies
978-1-118-49123-2

## Music

Blues Harmonica
For Dummies
978-1-118-25269-7

Guitar For Dummies,
3rd Edition
978-1-118-11554-1

iPod & iTunes
For Dummies, 10th Edition
978-1-118-50864-0

## Programming

Beginning Programming
with C For Dummies
978-1-118-73763-7

Excel VBA Programming
For Dummies, 3rd Edition
978-1-118-49037-2

Java For Dummies,
6th Edition
978-1-118-40780-6

## Religion & Inspiration

The Bible For Dummies
978-0-7645-5296-0

Buddhism For Dummies,
2nd Edition
978-1-118-02379-2

Catholicism For Dummies,
2nd Edition
978-1-118-07778-8

## Self-Help & Relationships

Beating Sugar Addiction
For Dummies
978-1-118-54645-1

Meditation For Dummies,
3rd Edition
978-1-118-29144-3

## Seniors

Laptops For Seniors
For Dummies, 3rd Edition
978-1-118-71105-7

Computers For Seniors
For Dummies, 3rd Edition
978-1-118-11553-4

iPad For Seniors
For Dummies, 6th Edition
978-1-118-72826-0

Social Security
For Dummies
978-1-118-20573-0

## Smartphones & Tablets

Android Phones
For Dummies, 2nd Edition
978-1-118-72030-1

Nexus Tablets
For Dummies
978-1-118-77243-0

Samsung Galaxy S 4
For Dummies
978-1-118-64222-1

Samsung Galaxy Tabs
For Dummies
978-1-118-77294-2

## Test Prep

ACT For Dummies,
5th Edition
978-1-118-01259-8

ASVAB For Dummies,
3rd Edition
978-0-470-63760-9

GRE For Dummies,
7th Edition
978-0-470-88921-3

Officer Candidate Tests
For Dummies
978-0-470-59876-4

Physician's Assistant Exam
For Dummies
978-1-118-11556-5

Series 7 Exam For Dummies
978-0-470-09932-2

## Windows 8

Windows 8.1 All-in-One
For Dummies
978-1-118-82087-2

Windows 8.1 For Dummies
978-1-118-82121-3

Windows 8.1 For Dummies,
Book + DVD Bundle
978-1-118-82107-7

 **Available in print and e-book formats.**

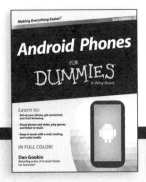   

Available wherever books are sold. **For more information or to order direct visit www.dummies.com**

# Take Dummies with you everywhere you go!

Whether you are excited about e-books, want more from the web, must have your mobile apps, or are swept up in social media, Dummies makes everything easier.

For Dummies is the global leader in the reference category and one of the most trusted and highly regarded brands in the world. No longer just focused on books, customers now have access to the For Dummies content they need in the format they want. Let us help you develop a solution that will fit your brand and help you connect with your customers.

## Advertising & Sponsorships

Connect with an engaged audience on a powerful multimedia site, and position your message alongside expert how-to content.

Targeted ads • Video • Email marketing • Microsites • Sweepstakes sponsorship

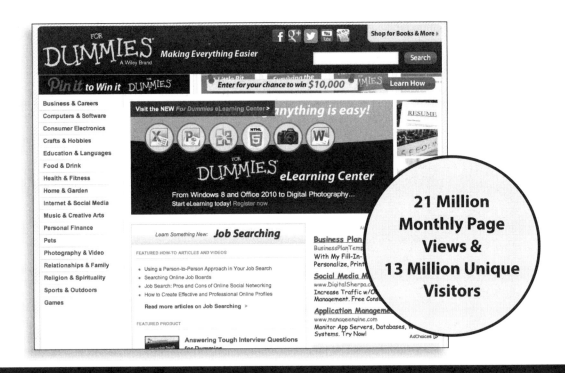

# Custom Publishing

Reach a global audience in any language by creating a solution that will differentiate you from competitors, amplify your message, and encourage customers to make a buying decision.

## Apps • Books • eBooks • Video • Audio • Webinars

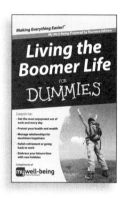

# Brand Licensing & Content

Leverage the strength of the world's most popular reference brand to reach new audiences and channels of distribution.

## For more information, visit www.Dummies.com/biz

A Wiley Brand

# Dummies products make life easier!

- DIY
- Consumer Electronics
- Crafts
- Software
- Cookware
- Hobbies
- Videos
- Music
- Games
- and More!

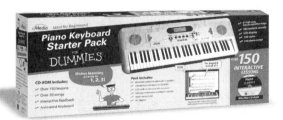

For more information, go to **Dummies.com** and search the store by category.

FOR DUMMIES
A Wiley Brand